Quṭb al-Dīn Shīrāzī and the Configuration of the Heavens

Archimedes

NEW STUDIES IN THE HISTORY AND PHILOSOPHY OF SCIENCE AND TECHNOLOGY

VOLUME 35

Archimedes has three fundamental goals; to further the integration of the histories of science and technology with one another: to investigate the technical, social and practical histories of specific developments in science and technology; and finally, where possible and desirable, to bring the histories of science and technology into closer contact with the philosophy of science. To these ends, each volume will have its own theme and title and will be planned by one or more members of the Advisory Board in consultation with the editor. Although the volumes have specific themes, the series itself will not be limited to one or even to a few particular areas. Its subjects include any of the sciences, ranging from biology through physics, all aspects of technology, broadly construed, as well as historically-engaged philosophy of science or technology. Taken as a whole, *Archimedes* will be of interest to historians, philosophers, and scientists, as well as to those in business and industry who seek to understand how science and industry have come to be so strongly linked.

For further volumes:
http://www.springer.com/series/5644

Kaveh Niazi

Quṭb al-Dīn Shīrāzī and the Configuration of the Heavens

A Comparison of Texts and Models

Springer

Kaveh Niazi
Berkeley
CA, USA

ISSN 1385-0180
ISBN 978-94-007-6998-4 ISBN 978-94-007-6999-1 (eBook)
DOI 10.1007/978-94-007-6999-1
Springer Dordrecht Heidelberg New York London

Library of Congress Control Number: 2013948472

Printed on acid-free paper

Springer is part of Springer Science+Business Media (www.springer.com)

Dedicated to my parents,
Mansour and Fereshteh Niazi,
for their loving support,
for instilling in us
a thirst for knowledge,
and for bestowing an abiding love for the
culture and history of Iran.

پس از هر دانستن دانستنی دیگرست ... و خداست جل جلاله که هیچ چیز بر وی پوشیده نیست.

Note on Transliteration

The historical sources used in writing this book were primarily in Arabic and Persian. Though it has not always been possible to be consistent with the transliteration scheme used in rendering these languages into English, I have generally tried to follow the scheme used by the *Encylopaedia of Islam* (using q for the letter ق, however). With few exceptions proper names have been rendered with their full battery of diacritics as these appear in the Persian and Arabic sources; so Khurāsān, rather than Khurasan or Khorasan. The idea has been to provide an easy to read text easy to read for readers that may not be familiar with Arabic or Persian. Proper names for which modern equivalents exist and that would have been changed beyond recognition (e.g., Konya and Aleppo) are left in their most recognizable form. As far as the astronomical nomenclature I have followed Prof. Ragep's edition of Ṭūsī's *al-Tadhkira*.

Acknowledgement

This work, which is based largely on my doctoral dissertation, could not have been completed without the help of many friends and mentors. First and foremost among these is my academic adviser George Saliba, to whom I must acknowledge an immense debt of gratitude. This work could not have been accomplished without his constant, kind, and generous support. I would also like to thank the other members of my thesis committee: Hamid Dabashi, Matthew Jones, Joel Kaye, and Nader Sohrabi for their unstinting encouragement, generosity, and support while I completed this work. John Walbridge provided invaluable guidance on biographical material related to Shīrāzī, for which I am immensely grateful. I owe as well a considerable debt of gratitude to F. Jamil Ragep for his research and especially for his edition of Ṭūsī's *al-Tadhkira*. Robert and Barbara Hanning, James Harney, Hasan Javadi, Hossein Kamaly, Abigail Schade each read parts of this manuscript and offered countless corrections and many helpful suggestions. I cannot adequately express my gratitude to them. Though the support of each of the listed persons was invaluable in the completion of this work, the responsibility for the many shortcomings and errors that remain in the work is exclusively mine.

When my research was just beginning, I was provided – through the generous support of then IRI ambassador to the United Nations, His Excellency J. Zarif – several key manuscripts that greatly facilitated my research, for which I would like to use this opportunity to again offer my most sincere and heartfelt thanks. Thanks also to Mr. Asgariyazdi and Mr. Abhari and their colleagues at the Majlis Library in Tehran, who expertly aided me with this project.

To my family I can only offer my humble gratitude for the love, joy and support that they have provided me throughout the various stages of preparing this book.

Acknowledgment

Contents

Chapter 1
Purpose and Background of Study

It is not natural that the stars should either move themselves ... or that they should be carried along certain circles. But there must exist spheres, made of the fifth essence, situated in the depth of the universe and moving there, some higher up, some arranged below them, some larger, some smaller, some hollow and some massive within the hollow ones, to which the planets are fastened in the manner of the fixed stars.

Theon of Smyrna, c. 120 C.E.

The great Plato, my friend, expects the true philosopher to take his mind from the perceptible and the totality of changing matter and to transfer astronomy beyond the heavens, to behold there absolute slowness and absolute speed with their true values. From these marvelous sights you seem to lead us down to those orbits in the heaven and to the observations of those practical people, the astronomers, and to those hypotheses which they have artificially devised on the grounds of their observations and which people like Aristarchus, Hipparchus, Ptolemy and others of their calibre used to din into our ears.

Proclus Diadochus, 410–485 C.E.

And it is necessary that motions that appear non-uniform rest upon that which entails their uniformity. Thus for each motion that is non-uniform, its corresponding angles or arcs in a given time period are compounds. So, if these principles are necessary, it is imperative at the same time for each planet to have several orbs due to the non-uniform motion it exhibits.

Quṭb al-Dīn Shīrāzī, 1235–1311 C.E. (634–710 A.H.)

1.1 Introduction

The goal of this study is a better understanding of the developments in astronomy in Persia in the late thirteenth and fourteenth-century, through a survey of the astronomical works of the polymath Quṭb al-Dīn Shīrāzī. Quṭb al-Dīn's books on astronomy include works in Arabic, the lingua franca of science in the Islamic world, as well as in his native language, Persian. Though in Shīrāzī's day, the era

K. Niazi, *Quṭb al-Dīn Shīrāzī and the Configuration of the Heavens: A Comparison of Texts and Models*, Archimedes 35, DOI 10.1007/978-94-007-6999-1_1,

of the Ilkhanid dynasty, Persian had come into its own as a sophisticated and supple vehicle for the production of literary and historical works, its use in scientific texts was considerably less common. Three of Shīrāzī's major works on astronomy – two in Arabic, one in Persian, written in the same stage of the author's life – provide, therefore, a rare opportunity to study the cultural interplay between the choice of language and the content of scientific works during this period. The works in question form the primary texts for this study, and all had, as their principal concern, the configuration of the celestial orbs, or *hay'at al-aflāk* in Arabic.

In Arabic the term *hay'a* denotes form or configuration, and the genre of astronomical writing to which it was applied aimed at a physically coherent description of the configuration of the universe as a set of nested spheres of specified dimensions subject to the laws of natural philosophy. This genre, which appeared at least as early as the eleventh century of the common era, does not have a precise analogue in the Greek tradition. Rather, *hay'a* grew out of Greek astronomy and inherited from it a long-standing debate with regard to the epistemological truths of astronomical knowledge.[1] How did the mathematical models that were used to predict the position of the celestial bodies correspond to reality? What was the nature of the celestial orbs, with which so many celestial observations could be described phenomenologically? How did these orbs interact with each other and with the heavenly bodies which appeared to be affixed to them? Given their success in describing the motion of the planets (even if, at times, this description was merely qualitative – as in providing a conceptual framework for the treatment of the retrograde motion of planets), how closely did the mathematical models of the astronomers correspond to the laws of natural philosophy? The impetus for *hay'a* research was the encounter of the scientists of the Islamic world with this Greek astronomical tradition, and the desire to combine a descriptive or geometrical astronomy, that was focused on the accurate prediction of the location of the celestial bodies, with a physics that aimed to describe the nature of the celestial bodies and their behavior.

The first concern, the development of precise predictive models, is exemplified by Ptolemy's monumental work on Astronomy, the *Almagest.* After its publication in the second century of the Common Era, this book was to serve as the main reference for astronomers in both the Hellenistic and Islamic traditions for the subsequent 14 centuries. Though resting on the Greek tradition of cosmology, as exemplified by Aristotle's *Metaphysics* and *De caelo*, Ptolemy's focus in much of the work is on the development of detailed mathematical models for the motions of the planets.[2] In the *Almagest* the mathematical models presented by

[1] The quotes on the previous page are from Samuel Sambursky, *The Physical World of Late Antiquity* (London: Routledge & Kegan Paul, 1987), 136; and Quṭb al-Dīn Shīrāzī, *Durrat al-tāj li-ghurrat al-dabāj* (Tehran: Chāpkhāneh-i Majlis, 1320), pt. 4, 67.

[2] George Saliba, "Aristotelian Cosmology and Arabic Astronomy," in *De Zénon d'Elée à Poincaré: Recueil d'études en hommage à Roshdi Rashed* (Louvain: Peeters, 2004), 254; Olaf Pedersen, *A Survey of the Almagest* (Odense: Odense Universitetsforlag, 1974), 122.

Ptolemy are informed by the cosmological views of the period, yet the physical or cosmological considerations remain in the background. For instance, while serving as the epistemic underpinnings of Ptolemy's astronomical theory, the spherical orbs – which are the purported movers of the planets – are barely mentioned in the *Almagest* at all.[3]

The second tradition of Hellenistic astronomy that was a source of the subsequent *hay'a* literature in the medieval period is represented by Ptolemy's *Planetary Hypotheses*, which was written after the *Almagest* and is considerably shorter. In the *Planetary Hypotheses* Ptolemy states that his aim is to treat the celestial motions in a more general way than he has in *The Almagest*, and in a manner, in his words, which "appeals more to the imagination."[4] Ptolemy's usage of the term "hypothesis" in the title of this work is distinct from the modern usage, and is a clue to his conceptualization of the book. Today a hypothesis means something akin to an untested theory, whereas Ptolemy used this word to mean a "system of explanation" or model.[5] In both the *Almagest* and the *Planetary Hypotheses* this term refers to mathematical as well as physical models akin to equatoria, i.e., devices constructed of wood or metal, used to depict the motion of the planets.[6] Ptolemy's focus in the *Planetary Hypotheses* is on providing a coherent depiction of the planets and the planetary orbs as physical objects. Thus, the challenges that would have faced him in the composition of the *Almagest* (namely, the need to carry out his theoretical work within the framework of Aristotelian cosmology) could only have been present to an even greater extent during the composition of the *Planetary Hypotheses*, once the physical nature of the planets and their orbs was taken into account.[7]

As far as his equatoria and the issues facing them, Ptolemy introduces a discussion of their limitations even in the *Almagest*. Invoking the perfection of the celestial realm in contrast to the imperfection of his equatoria Ptolemy states:

> Now let no one, considering the complicated nature of our devices, judge such hypotheses to be over-elaborated. For it is not appropriate to compare human [constructions] with divine, nor to form ones beliefs about such great things on the basis of very dissimilar analogies.... Rather, one should try, as far as possible, to fit the simpler hypothesis to the heavenly motions, but if this does not succeed one should apply hypotheses which do fit. For provided that each of the phenomena is duly saved by the hypotheses, why should

---

[3]Pedersen, *A Survey of the Almagest*, 34; Naṣīr al-Dīn Muḥammad ibn Muḥammad Ṭūsī, *Naṣīr al-Dīn al-Ṭūsī's Memoir on Astronomy = al-Tadhkira fī 'ilm al-hay'a.* F. Jamil Ragep, Ed. (New York: Springer-Verlag, 1993), 25; Sambursky, *The Physical World of Late Antiquity*, 140; Bernard R. Goldstein, "The Arabic Version of Ptolemy's Planetary Hypotheses," *Transactions of the American Philosophical Society* 57, no. 4, New Series (1967): 3.

[4]Pedersen, *A Survey of the Almagest*, 392; Goldstein, "The Arabic Version of Ptolemy's Planetary Hypotheses," 3.

[5]Ptolemy, *The Almagest* (Princeton: Princeton University Press, 1998), 23.

[6]N.M. Swerdlow, *Mathematical Astronomy in Copernicus's De Revolutionibus* (New York: Springer-Verlag, 1984), 40.

[7]Saliba, "Aristotelian Cosmology and Arabic Astronomy," 254; Goldstein, "The Arabic Version of Ptolemy's Planetary Hypotheses," 39.

> anyone think it strange that such complications can characterize the motions of the heavens when their nature is such as to afford no hindrance, but of a kind to yield and give way to the natural motions of each part.[8]

Among other things Ptolemy appears to argue here for a distinction between the celestial and terrestrial realms, i.e., the realm of the heavenly orbs and the realm of mechanical devices meant to approximate their behavior. A good deal of the effort of the scientists of the Islamic world was exerted on providing theoretical formulations of the motion of the planets that would describe the observable phenomena and yet be free of the sort of complications that Ptolemy alluded to.[9]

Aristotle, who was responsible for articulating the cosmological system within which Ptolemy was to carry out his work, was not concerned with producing detailed models of planetary motion.[10] Where he does describe the intricacies of the celestial models under consideration his remarks are qualitative. In the *Metaphysics* he writes:

> Eudoxus held that the motion of the Sun or of the Moon involves, in either case, three spheres, of which the outermost is the sphere of the fixed stars, and the second revolves in the circle which bisects the zodiac, and the third in the circle which is inclined across the breath of the zodiac....and he held that the motion of the planets involves, in each case, four spheres, and of these also the first and second are the same as before . . . the third sphere of all planets has its poles in the circle which bisects the zodiac, and the fourth sphere moves in the circle inclined to the equator of the third . . . and the number of all the spheres – those which move the planets and those which counteract these – will be fifty-five.[11]

Ragep discusses how the elaborate system of counter-rolling spheres described in this fragment indicates Aristotle's concern with the physical nature of the spheres in question.[12] Certainly, here, as elsewhere in the cosmological sections of *De caelo* and the *Metaphysics*, the description is merely qualitative. Aristotle's goal was clearly not to provide models with accurate predictive abilities. Instead, it fell on Ptolemy to propose a collection of detailed mathematical models of the universe, while remaining within the constraints of Aristotelian cosmology. This cosmology imposed a strict set of requirements on celestial motion – the requirement of uniform circular motion that was centered on the earth, being a primary example.[13]

In the *Planetary Hypotheses*, Ptolemy not only lists the distances and sizes of the planets, but he includes, as well, a description of what was to become the basic conceptual framework for *hay'a*, *i.e.*, a scheme in which the spheres of adjacent planets are nested so that the greatest distance of a given planet relative to the center

---

[8]Ptolemy, *The Almagest*, 600.

[9]George Saliba, *Islamic Science and the Making of the European Renaissance* (Cambridge, Mass: MIT Press, 2007), 106.

[10]Saliba, "Aristotelian Cosmology and Arabic Astronomy," 253.

[11]Aristotle, *The Metaphysics* (Cambridge, Mass: Harvard University Press, 1956); Ṭūsī, *Naṣīr al-Dīn al-Ṭūsī's Memoir*, 26.

[12]Ṭūsī, *Naṣīr al-Dīn al-Ṭūsī's Memoir*, 26.

[13]Pedersen, *A Survey of the Almagest*, 34; Saliba, "Aristotelian Cosmology and Arabic Astronomy," 253.

of the world is equal to the least distance of the next planet farther from the center of the world, and so on, all the way to the sphere of the fixed stars.[14]

Given the fact that the purported orbits of the heavenly bodies were, according to the universally-held Aristotelian belief system, circular and geocentric (rather than elliptical and heliocentric as we know them today to be), it was necessary for Ptolemy to devise ingenious mathematical formulations that were physically unrealizable, in order to account for the variable velocity of the planets in their orbits. In the *Almagest* the deferent spheres for the Moon and the planets, for instance, are formulated to "rotate" about a point that is not coincident with their axes. While admissible as a mathematical feature of the theory when treating the orbs of the planets in an abstract and mathematical sense, this element of Ptolemy's theory was physically untenable as far as the authors of the *hay'a* tradition were concerned. The issue, the so-called "equant problem," is not raised by Ptolemy in this work but was one of the main driving forces for the theoretical work of the *hay'a* authors, as discussed by Saliba and others.[15]

At its roots the problem of the equant is the problem of reconciling detailed and descriptive mathematical models of the motion of the celestial bodies with a coherent physical picture. Though, as we have seen, this was already a preoccupation of the astronomers of the Hellenistic period, the attempt to bring these two conceptions of astronomy into agreement proved especially productive for the scientists of the Islamic world during the medieval era.[16]

The desire to arrive at an astronomical formulation that was in accord with physics, can be seen, for example in two early *hay'a* works by Ibn al-Haytham, known to the Latin west as Alhazen (c. 965–1039 C.E., c. 354–430 A.H.): *al-Maqāla fī hay'at al-'ālam* (or "Treatise on the Configuration of the Universe," henceforth the *Maqāla*), and *al-Shukūk 'alā Baṭlamyūs*, (or "Doubts Concerning Ptolemy").[17] In his earlier work, the *Maqāla*, one of Ibn al-Haytham's goals appears to have been a re-rendering of the Ptolemaic system with an emphasis on the spherical orbs of the heavens as three-dimensional bodies.[18] This is made clear with a statement regarding the limitations of the mathematical formulations within the Ptolemaic tradition:

> Since those theories [of Ptolemy], that is, those which point to the form of the figure and the laws of the motions by means of proper observation and correct proofs are, however, based upon the motions of imaginary points on the circumferences of intellected circles according to what is demonstrated in those books of theirs which we have; and, likewise, [those points]

[14]Goldstein, "The Arabic Version of Ptolemy's Planetary Hypotheses," 7.

[15]Ptolemy, *The Almagest*, 422; Swerdlow, *Mathematical Astronomy in Copernicus's De Revolutionibus*, 40.

[16]Saliba, *Islamic Science and the Making of the European Renaissance*, 132–170; Sambursky, *The Physical World of Late Antiquity*, 133–145.

[17]George Saliba, *A History of Arabic Astronomy: Planetary Theories During the Golden Age of Islam* (New York: New York University Press, 1994), 13; Ṭūsī, *Naṣīr al-Dīn al-Ṭūsī's Memoir*, 49.

[18]V. Minorsky, "Sulṭānīya," *Encyclopaedia of Islam, Second Edition* (Brill Online, 2010), http://www.brillonline.nl/subscriber/entry?entry=islam_COM-1118.

> are assigned by indication on their part, but not explicitly, to the surfaces of solid spheres which, in fact, are the things which have those motions on those points, it turned out that their theory insofar as they explained it was limited to those circles and points only.[19]

If the orbs in which the planets are embedded are said to be three dimensional objects, Ibn al-Haytham appears to be saying, then a proper treatment of the motion of a planet must include the reality of these spheres in its derivation (and not be limited, as Ptolemy's had been in the *Almagest*, to treating cross-sections of spheres on a planar surface). Ibn al-Haytham continues in the *Maqāla*: "Since our doctrine is in accordance with what he [Ptolemy] explained and arranged, and he avoided the use of any bodies, we investigated each of the motions ... in such a manner that that motion may appear to be the result of a spherical body that is moving with a simple, continuous, and unceasing motion." This declaration provides a statement of purpose for the composition of the *Maqāla* while acknowledging its debt to Ptolemy, as well.[20] Interestingly, in Ibn al-Haytham's later work on *hay'a*, *al-Shukūk 'alā Baṭlamyūs*, such accord with Ptolemaic theory appears no longer to have been tenable for the author. As the title suggests, this work is a critique of the physical inconsistencies of Ptolemaic theory as they appear in both the *Almagest* and the *Planetary Hypotheses*.[21] These physical inconsistencies were caused by the fact that the celestial orbs were constrained, by prevailing Aristotelian notions of how the universe worked, to move with a uniform angular velocity, as we saw before. Using this as a criterion, the list of non-physical elements in Ptolemy's theory that Ibn al-Haytham identified includes the irregular rotation of spheres (as in the case of the motion of the deferent sphere for the planets and its posited "uniformity" about a point distinct from the center of the sphere), back-and-forth (i.e., non-circular) motions of the lunar epicycle,[22] and oscillations of orbital planes to account for the latitudes of the planets. The non-physical features of Ptolemaic astronomy were to preoccupy the scientists of the *hay'a* tradition of the ensuing centuries.

Writing in the thirteenth century – in what was a period of efflorescence for *hay'a* research – renowned and savant Naṣīr al-Dīn Ṭūsī (1201–1274 C.E., 597–672 A.H.) includes sixteen objections to Ptolemy in his *hay'a* work *al-Tadhkira fī 'ilm al-hay'a*, or *Memoir on Astronomy* (henceforth referred to as the *Tadhkira*). Ṭūsī's objections include: the irregular motion of the deferents of the Moon, Mercury, Venus, Mars, Jupiter and Saturn; latitudinal deviation and latitudinal slant for Venus and Mercury; the oscillation of the equators of the deferent orbs for Venus and Mercury; and the back and forth oscillation of the lunar epicycle.[23]

---

[19] Abū ʿAlī al-Ḥasan Ibn al-Haytham, *Ibn al-Haytham's On the Configuration of the World*, Tzvi Langermann, Ed. (New York: Garland, 1990), 53.

[20] Ibn al-Haytham, *Ibn al-Haytham's On the Configuration of the World*, 53.

[21] Saliba, *Islamic Science and the Making of the European Renaissance*, 97–106.

[22] In Ptolemy's scheme the lunar epicycle was the orb which carried the Moon. The epicycle in turn was carried by the deferent orb.

[23] Ṭūsī, *Naṣīr al-Dīn al-Ṭūsī's Memoir*, 50. A comparison of the two lists demonstrates Ṭūsī's debt to Ibn al-Haytham, whom he mentions repeatedly in his works.

Ṭūsī himself provides a solution to the first set of issues, i.e., the equant or the irregular motion for all the planets (save the Sun and Mercury) by relying on a mathematical formulation now referred to as the Ṭūsī Couple.[24] Mu'ayyad al-Dīn al-'Urḍī (d. 1266 C.E.), perhaps the most innovative astronomer of his era, was to provide another original solution to the problem of the equant, one that was based on a mathematical theorem now known as 'Urḍī's Lemma.[25] As Ṭūsī's student Quṭb al-Dīn Shīrāzī, who is the subject of the present study, was able to rely on the works of his predecessors Ṭūsī and al-'Urḍī, and, by incorporating the Ṭūsī couple and 'Urḍī's Lemma, to propose additional planetary models of increasing complexity. Both of these mathematical formulations appear as well in the works of Copernicus, thus linking the astronomy of the Early Modern Period in Europe to the research of al-'Urḍī, Ṭūsī, and their fellow astronomers in the Islamic world.[26]

1.2 Shīrāzī and His Era

1.2.1 The Marāgha School

The term Marāgha School was first coined by Kennedy, and used in a 1966 article in reference to the group of astronomers mentioned above, i.e., Ṭūsī, al-'Urḍī, Ṭūsī, Shīrāzī, and others who were active in thirteenth-century Ilkhanid Iran.[27] Also included in the grouping were scientists such as the Damascene astronomer Ibn al-Shāṭir (1304–1375 C.E., 704–777 A.H.) whose theoretical work can be viewed as a continuation of that of the aforementioned scientists.[28] The term Marāgha refers to the site of the great observatory commissioned by the grandson of Chingiz Khan, Hülegü (or Hulāgū, as he is referred to in the Islamic world), who appears to have settled in the city in 1258 C.E. after the fall of Baghdād. At roughly the same time Ṭūsī selected a site near the city for the construction of the observatory.[29] Though

---

[24] Saliba, *Islamic Science and the Making of the European Renaissance*, 155.

[25] Saliba, *Islamic Science and the Making of the European Renaissance*, 151.

[26] Swerdlow, *Mathematical Astronomy in Copernicus's De Revolutionibus*, 47.

[27] E.S. Kennedy, "Late Medieval Planetary Theory," *Isis* 57, no. 3 (Autumn 1966): 365; Mu'ayyad al-Dīn Ibn Burayk Urḍī al-, *Kitāb al-hay'a*, George Saliba, Ed. (Bayrūt: Markaz Dirāsāt al-Waḥdah al-'Arabīyah, 1990), 29.

[28] Kennedy, "Late Medieval Planetary Theory," 365.

[29] Rashīd al-Dīn Ṭabīb, *Jāmi' al-tawārīkh*, Bahman Karimi, Ed. (Tehrān: Iqbāl, 1338), 717–718; Aydın Sayılı, *The Observatory in Islam and Its Place in the General History of the Observatory* (Ankara: Türk Tarih Kurumu Basımevi, 1960), 189–234; John Andrew Boyle, "Dynastic and Political History of the Īl-khāns," in *The Cambridge History of Iran*, vol. 5, Ed. J.A. Boyle (Cambridge: Cambridge University Press, 1968), 349; Rashīd al-Dīn Ṭabīb, *Rashiduddin Fazlullah's Jami'u't-Tawarikh = Compendium of Chronicles* (Cambridge, Mass.: Harvard University, Dept. of Near Eastern Languages and Civilizations, 1998), 717–718; J. Samsó, "Marṣad," *Encyclopaedia of Islam, Second Edition* (Brill Online, 2010), http://www.brillonline.nl/subscriber/entry?entry=islam_SIM-4972.

the term Marāgha school is perhaps useful in identifying a shared approach in addressing the issues facing Ptolemaic astronomy, it can also be misleading due to its lack of precision. As we have seen, for example, not all astronomers grouped in the Marāgha school actually lived there.[30]

Funded by *awqāf* revenues, the Marāgha observatory was to continue its operation for more than 50 years.[31] The first director of the observatory was Naṣīr al-Dīn Ṭūsī himself, who staffed the observatory with astronomers from as far afield as China.[32] Al-'Urḍī's name has been preserved as the builder of the scientific instrumentation at the observatory.[33] Though there is no documented evidence that Shīrāzī worked at the observatory, his tutelage under Ṭūsī and his close association with him make an association with the observatory, in some form, highly probable.[34]

The transmission of the intellectual tradition of the Marāgha school to Early Modern Europe has been an area of active research. A considerable amount of evidence confirms that this transmission did indeed occur. The list of models within Copernicus's *De revolutionibus* and *Commentariolus* that can be traced to the aforementioned astronomers includes those devised by Ṭūsī, al-'Urḍī, and Ibn al-Shāṭir.[35] In addition there is conclusive evidence for knowledge of the planetary theory of Ṭūsī in Italy in the early sixteenth century.[36] Though the precise path for the transmission of this knowledge to Copernicus has yet to be determined, it is likely that he learned of it himself during his stay in Padua during the years 1501–1503 C.E.; perhaps through a work, similar to one of Shīrāzī's books on *hay'a*, that included references to a collection of techniques devised by Ṭūsī, al-'Urḍī and others.[37]

---

[30]For this reason, the use of this term will be avoided here. Swerdlow, *Mathematical Astronomy in Copernicus's De Revolutionibus*, 295; George Saliba, "The First Non-Ptolemaic Astronomy at the Maraghah School," *Isis* 70, no. 4 (December 1979): 571–576; The same is true for Ṭūsi's theoretical work on the motion of the planets; Ṭūsī, *Naṣīr al-Dīn al-Ṭūsī's Memoir*, 14.

[31]Sayılı, *The Observatory in Islam and Its Place in the General History of the Observatory*, 207–211; Ṭūsī, *Naṣīr al-Dīn al-Ṭūsī's Memoir*, 14. *Awqāf* refers to the system of religious endowments in the Islamic world.

[32]Ṭūsī, *Naṣīr al-Dīn al-Ṭūsī's Memoir*, 14; Willy Hartner, "The Astronomical Instruments of Cha-ma-lu-ting, Their Identification, and Their Relations to the Instruments of the Observatory of Marāgha‡," *Isis* 41, no. 2 (July 1950): 184–194.

[33]Urḍī, *Kitāb al-hay'a*, 30.

[34]Muhammad Mudarris Razavi, *Aḥwāl wa Athār-i Muḥammad Ibn Muḥammad Ibn al-Ḥasan al-Ṭūsī* (Tehran: Intisharat-i daneshgah-i Tehran, 1955), 30.

[35]Swerdlow, *Mathematical Astronomy in Copernicus's De Revolutionibus*, 47; Otto Neugebauer, *The Exact Sciences in Antiquity.* (Providence: Brown University Press, 1957), 203; Kennedy, "Late Medieval Planetary Theory," 365; N.M. Swerdlow, "The Derivation and First Draft of Copernicus's Planetary Theory: A Translation of the Commentariolus with Commentary," *Proceedings of the American Philosophical Society* 117, no. 6 (December 31, 1973): 500; Saliba, *Islamic Science and the Making of the European Renaissance*, 206–209.

[36]Swerdlow, *Mathematical Astronomy in Copernicus's De Revolutionibus*, 48.

[37]Swerdlow, *Mathematical Astronomy in Copernicus's De Revolutionibus*, 48; Saliba, *Islamic Science and the Making of the European Renaissance*, 212.

1.2.2 Shīrāzī: Preliminary Remarks

Described as "one of the greatest Persian scientists of all times," and "one of the foremost thinkers and scholars of Islam," Quṭb al-Dīn Shīrāzī (1235–1311 C.E., 634–710 A.H.) was, much like his teacher Ṭūsī, a polymath who wrote on astronomy, philosophy, theology, and medicine.[38] He is best remembered today for his commentary on the *Philosophy of Illumination* by the Persian illuminationist philosopher Suhrawardī.[39] A large encyclopedic work of his, *Durrat al-tāj li-ghurrat al-Dabāj* (or "the Pearl in the Crown for the Brow of al-Dabāj," henceforth, the *Durra*) is well known, though today it is studied primarily as a work of Persian literature.[40] The dearth of published works by Shīrāzī, noted by Nasr in 1976, is strangely at odds with Shīrāzī's reputation and has not improved substantially since then.[41]

Three of Shīrāzī's texts on astronomy will be examined in some detail in Chap. 4. These are (1) *Nihāyat al-idrāk fī dirāyat al-aflāk* ("The Limits of Attainment in the Understanding of the Heavens," henceforth the *Nihāya*) which is the earliest of Shīrāzī's major works on *hay'a*, (2) *al-Tuḥfa al-shāhīya fī 'ilm al-hay'a* ("The Royal Offering Regarding the Knowledge of the Configuration of the Heavens," henceforth the *Tuḥfa*) a shorter work written less than 4 years later, and (3) *Ikhtīyārāt-i Muẓaffarī*, a *hay'a* text in Persian. The title of *Ikhtīyārāt-i Muẓaffarī* indicates that it is an astrological work dedicated to Shīrāzī's patron, Muẓaffar al-Dīn, and this book will henceforth be referred to as the *Ikhtīyārāt*.[42] While portions of the *Nihāya* and the *Tuḥfa* have been the subject of articles by historians of science, including Kennedy, Saliba, and others, none of these works has been edited or extensively translated. The goal of this thesis is to contribute in a small way, therefore, to the scholarship concerning an overlooked medieval scientist, by further exploring the development Shīrāzī's thoughts on astronomy as exhibited in these three closely-related works, and by an examination of the social and cultural influences on Shīrāzī as they are manifested by his choice of language.

---

[38]George Sarton, *Introduction to the History of Science* ... (Baltimore, Pub. for the Carnegie Institution of Washington, by the Williams & Wilkins Co., 1962), 1017; S.H. Nasr, *The Islamic Intellectual Tradition in Persia* (Richmond, Surrey [England]: Curzon Press, 1996), 217.

[39]See John Walbridge, *The Science of Mystic Lights: Quṭb al-Dīn Shīrāzī and the Illuminationist Tradition in Islamic Philosophy* (Cambridge, Mass: Distributed for the Center for Middle Eastern Studies of Harvard University by Harvard University Press, 1992).

[40]Shīrāzī, *Durrat al-tāj li-ghurrat al-dabāj*; Walbridge, *The Science of Mystic Lights*, 176. Variant spellings of the name of the dedicatee are Dobāj, and Dabbāj.

[41]Nasr, *The Islamic Intellectual Tradition in Persia*, 217.

[42]The word *Ikhtīyārāt* means choices or selections in Persian. It refers to an astrological genre which was focused on determining the auspiciousness of a given day for a given action. Ṭūsī appears to have contributed to this genre, as well. See David Pingree, "EḴTĪĀRĀT," *Encyclopaedia Iranica*, December 15, 1998, http://www.iranica.com/articles/ektiarat. Despite its name, the *Ikhtīyārāt* belongs to the *hay'a* genre.

1.2.3 Status of Scholarship on Shīrāzī's Astronomy

One of the earliest discussions of Shīrāzī's relevance to the history of astronomy appears in Kennedy's 1966 article referred to above. In this article a mathematical formulation known as the principle (or the hypothesis) of the "maintainer and the director" (*aṣl al-ḥafiẓa wa al-mudīr*) is ascribed to Shīrāzī.[43] In the same article, however, Kennedy notes textual clues within the *Nihāya*, such as Shīrāzī's allusion to the "master of this method" (as "one of the foremost practitioners of this science [i.e., *hay'a*]") as an indication that the formulation may have originated with someone else.[44] Indeed, as we shall see, there are several other references within Shīrāzī's works to the "master of the principle of the maintainer and the director."[45] In a series of articles published in the late 1970s Saliba demonstrated that the astronomer Shīrāzī was referring to as "the master of this method" was none other than al-'Urḍī himself.[46] Shīrāzī's reluctance in identifying those of his immediate predecessors such as al-'Urḍī and Ṭūsī upon whose work he relies heavily for his astronomical works is rather puzzling. It is worth noting here, however, that in the *Durra* Shīrāzī praises al-'Urḍī explicitly in a passage on planetary motions:

> And the science that is devoted to [the motion of the planets] is *hay'a*, numerous and many-branched are its various subjects. And it is of those excellent sciences that offer proof as to the grandeur of the Creator, may He be glorified. And the knowledgeable savant Mu'ayyad al-Dīn al-'Urḍī has studied it in a manner such that no one else has surpassed him in it.[47]

In addition to the research carried out on Shīrāzī's work by Kennedy, and Saliba, one of the chapters of the *Tuḥfa* has been edited and translated by Morrison.[48] This chapter includes a systematic presentation of various mathematical formulations such as the Ṭūsī couple and 'Urḍī's Lemma, as well as others. We will have the opportunity to refer to this chapter in Chap. 4.

1.2.4 Methodology and Approach

The methodology for the study is suggested by Shīrāzī's works themselves. A textual comparison of the three works should help answer the following set

---

[43]Other references to this formulation translate the Arabic expression as the "maintainer and the dirigent," as well as the "protector and the dirigent." The significance of this formulation will be examined in Chap. 4.

[44]Shīrāzī, *Nihāyat al-idrāk fī dirāyat al-aflāk*, Köprülü MS 957, 82r.

[45]These references will be discussed in Chap. 4.

[46]Saliba, *A History of Arabic Astronomy*, 113–135.

[47]Shīrāzī, *Durrat al-tāj li-ghurrat al-dabāj*, pt. 4, 67.

[48]Robert Morrison, "Qutb al-Din al-Shirazi's Hypotheses for Celestial Motions," *Journal for the History of Arabic Science* 13 (2005).

of questions: How are the three works related? Why did Shīrāzī choose to write three books in two different languages covering presumably much of the same material? What were Shīrāzī's criteria for including or omitting material in his books on *hay'a*? What does a comparison of these three works tell us about Shīrāzī's approach to his astronomical research and his models for the planetary motions?

Another set of questions is suggested by the view of Shīrāzī as a scientist embedded in the society of late thirteenth-century Persia. Given the fact debt owed to his predecessors, how did Shīrāzī how did Shīrāzī incorporate and present the material of his predecessors in his work? How did issues of patronage affect the production of the works in question? Who were Shīrāzī's patrons and how did they end up listed as the dedicatees of Shīrāzī's astronomical works?

As is well known, Arabic has been the language of science *par excellence* in the Islamic world. A final set of questions, then, revolves around Shīrāzī's choice of language. Why did he write one of his major books on astronomy in Persian? How did this choice of language affect the content of these books? Did the decision to compose the *Ikhtīyārāt* in Persian influence or reflect, as has been suggested, the technical sophistication of the work?

The choice of the chapter on the superior planets is driven by the importance of this chapter in highlighting both Shīrāzī's technical capacities as a scientist as well as his relationship with his predecessors, upon whose work he solidly rests his. The technical nature of this chapter should allow, as well, for a careful examination of the use of Persian. It should be noted, however, that this study represents a mere beginning. It is to be hoped that the study of Shīrāzī's works on *hay'a* and on other topics will continue, culminating with edited translations of the works of this important thirteenth century C.E. figure.

1.2.5 Outline

In the remainder this introductory chapter (Sect. 1.3) I present the list of sources for this book. In Chap. 2 I sketch the historical backdrop to the era in which Shīrāzī lived, and in Chap. 3 present what has reached us in regard to Shīrāzī's life. Chapter 4 consists of a comparison of the chapters on the superior planets as they appear in each of Shīrāzī's *hay'a* books mentioned above. As Appendix A indicates, each of these books is organized in a nearly identical manner by being divided into four large sections. The section of primary interest for our study is the second section, which includes the planetary models for the Sun, Moon, and the superior and inferior planets. In this same section Shīrāzī includes a chapter on the mathematical "hypotheses" or principles (such as the Ṭūsī couple) on which he bases his subsequent discussion. This chapter yields a considerable amount of material pertinent to our discussion, and will be presented prior to our discussion of the Moon and the superior planets. Once the stage has been set for our discussion of the superior planets, these texts will be used to illustrate the development of

Shīrāzī's thought on the configuration of the planetary orbs, his use of language, and his manner of presentation. Chapter 5 will include a discussion of how the choice of language in these works is manifested in the content of each work. Chapter 6 will provide a summary of our findings together with some concluding remarks and observations.

1.3 The Sources

1.3.1 Astronomical

Shīrāzī's Books on Astronomy

The three books of Shīrāzī on astronomy that form the critical primary sources for this study were listed in Sect. 1.2.3 of this chapter. They are the *Nihāya*, and the *Tuḥfa* in Arabic, and the *Ikhtīyārāt* in Persian. The manuscripts that have reached us generally consist of 200 or more folios, and as noted none of these works has been edited.[49] The manuscript copies of these books that were used for this study will be described in Chap. 4.

The *Tadhkira* by Ṭūsī

This book is in Arabic and has been translated and edited by Ragep, who in so doing has created a comprehensive work of reference for the students of medieval astronomy. Ragep was able to identify two versions of the *Tadhkira*: what he called a Marāgha, and, a later, Baghdād version.[50] Interestingly some of the changes in the Baghdād version of the *Tadhkira* may have been due to Shīrāzī.[51] In addition Shīrāzī appears to have had his own personal copy of this work.[52]

Others of Ṭūsī's works on *hay'a* that have been published include *Zubdat al-idrāk fī hay'at al-aflāk*, or the "Essential Understanding of the Configuration of the Orbs," and *al-Risāla al-mu'īnīya*, or "The Mu'īnīya Epistle," and *Ḥall-i mushkilāt-i mu'īnīya*, or "The Solution of the Difficulties of the Mu'īnīya."[53] The first book

[49] A fourth work by Shīrāzī on astronomy, *Fa'altu fa lā talum*, is a polemical work of a later date and is not part of the present study.

[50] Ṭūsī, *Naṣīr al-Dīn al-Ṭūsī's Memoir*, 70–71.

[51] Ṭūsī, *Naṣīr al-Dīn al-Ṭūsī's Memoir*, 73–75.

[52] Ṭūsī, *Naṣīr al-Dīn al-Ṭūsī's Memoir*, 78.

[53] Naṣīr al-Dīn Muḥammad ibn Muḥammad Ṭūsī, *Zubdat al-idrāk fī hay'at al-aflāk: ma'a dirāsah li-manhaj al-Ṭūsī al-'ilmī fī majāl al-falak*, 1st ed. (al-Iskandarīyah: Dār al-Ma'rifah al-Jāmi'īyah, 1994); Naṣīr al-Dīn Muḥammad ibn Muḥammad Ṭūsī, *al-Risālah al-mu'īnīyah* (Tehran: Chāpkhānah-'i Dānishgāh, 1335); Naṣīr al-Dīn Muḥammad ibn Muḥammad Ṭūsī, *Ḥall-i mushkilāt-i mu'īnīyah* (Teheran: Chāpkhānah-'i Dānishgāh, 1335).

is in Arabic, and the other two are in Persian. With respect to these works Ragep states that the *Zubdat al-idrāk fī hay'at al-aflāk* appears to be a simplified work, and the latter two works (both in Persian) appear to be the predecessors of the *Tadhkira*.[54] *Al-Risāla al-muʿīnīya*, and *Ḥall-i mushkilāt-i muʿīnīya* have been the subject of several in-depth studies by Ragep.[55]

Al-'Urḍī's *Kitāb al-hay'a*

Kitāb al-hay'a or "The Book of *hay'a*" by al-'Urḍī was written before 1259 C.E.[56] In it al-'Urḍī presents the celebrated lemma that allowed him to deal with important contradictions in the Greek astronomical tradition (namely, those concerning the aforementioned equant problem). This work has been edited and published by Saliba. Only three manuscript copies of this work are extant.[57] That Shīrāzī knew 'Urḍī's work is clear from the fact that he relies on 'Urḍī's models in his own work.[58] Furthermore, that Shīrāzī knew 'Urḍī's *Kitāb al-hay'a* itself is evident by the presence in the *Nihāya* of an extended section, several paragraphs long, that intersperses direct quotes from 'Urḍī's work with paraphrased fragments, in addition to other paraphrased fragments from this work.[59]

1.3.2 Biographical

Shīrāzī's Autobiographical Notes

Shīrāzī himself wrote an autobiography in the introduction to his commentary on Avicenna's *Canon of Medicine*, *al-Tuḥfa al-sa'dīya fī al-ṭibb*.[60] This autobiography

---

[54] Ṭūsī, *Naṣīr al-Dīn al-Ṭūsī's Memoir*, 67.

[55] F.J. Ragep, "The Persian Context of the Ṭūsī Couple," in *Naṣīr al-Dīn al-Ṭūsī: Philosophe et Savant du XIIIe Siècle* (Tehran: Institut français de recherche en Iran/Presses universitaires d'Iran, 2000), 113–130; F.J. Ragep, "Ibn al-Haytham and Eudoxus: The Revival of Homocentric Modeling in Islam," in *Studies in the History of the Exact Sciences in Honour of David Pingree*, Charles Burnett, Jan P. Hogendijk, Kim Plofker and Michio Yano, Eds. (Leiden: E.J. Brill, 2004), 786–809. F.J. Ragep and Behnaz Hashemipour, "Juft-i Ṭūsī (the Ṭūsī Couple)," in *The Encyclopaedia of the World of Islam* (Tehran: Encyclopaedia Islamica Foundation, 2006), vol. X, pp. 472–475.

[56] Urḍī, *Kitāb al-hay'a*, 31.

[57] Urḍī, *Kitāb al-hay'a*, 8.

[58] Saliba, *A History of Arabic Astronomy*, 113–135.

[59] This occurs in the discussion immediately prior to a description of 'Urḍī's Lemma. The figure in Shīrāzī's book follows the lettering scheme of al-'Urḍī, as well. Quṭb al-Dīn Shīrāzī, *Nihāyat al-idrāk fī dirāyat al-aflāk*, Köprülü MS 957, 73v., and the *Kitāb al-hay'a*, Marsh MS 621, 158v. See also Saliba, *A History of Arabic Astronomy*, 131 and Urḍī, *Kitāb al-hay'a*, 222.

[60] Walbridge, *The Science of Mystic Lights*, 186.

covers in some detail, Shīrāzī's early training as a physician and some other events of Shīrāzī's life up to the accession of Ghāzān in 1295 C.E. In his autobiography Shīrāzī describes in some detail his upbringing in a medical family and the hardships endured in authoring his commentary on Avicenna's seminal work. Though Shīrāzī's commentary on *the Canon* has not been translated or studied in detail, it is, in what is perhaps a measure of the importance the author attached to it, his only known work that includes such autobiographical material.[61] This biographical note has been reproduced nearly in its entirety in the edition of the *Durra* that was used for this study.[62]

Biographical Dictionaries

A monumental work by Ibn al-Fuwaṭī (1244–1323 C.E./642–723 A.H.) who was a librarian at Marāgha (and who appears to have known Shīrāzī personally), the *Majma' al-ādāb fī mu'jam al-alqāb*, has only survived in an abridged form.[63] The biography of Shīrāzī that appears in the surviving work is not extensive, even though it is likely that information from the original work has found its way into other biographies. *Tārīkh 'ulamā' Baghdād, al-musammā muntakhab al-mukhtār*, a fourteenth century history by Muḥammad ibn Rāfi' al-Sallāmī (d. 1372 C.E./774 A.H.) has a fairly lengthy biography of Shīrāzī.[64] Ibn Ḥajar al-'Asqalānī (1372–1449 C.E./773–852 A.H.) also includes a substantial entry on Shīrāzī in his *al-Durar al-kāmina* with some material that is not found in either the abridged dictionary of Ibn al-Fuwaṭī or in the work of al-Sallāmī.[65] Other early sources of biographical information for Shīrāzī include *Ṭabaqāt al-shāfi'īya al-kubrā*, by Tāj

[61] Walbridge, *The Science of Mystic Lights*, 186.

[62] Shīrāzī, *Durrat al-tāj li-ghurrat al-dabāj*, 8–11.

[63] 'Abd al-Razzaq ibn Ahmad Ibn al-Fuwaṭi, *Majma' al-adāb fi mu'jam al-alqāb*, Kazim, M., Ed. (Tehran: Muassasāt al-ṭibā' a wa al-nashr, Wizārat al-thaqāfah wa al-irshād al-Islāmī, 1995); F. Rosenthal, "Ibn al- Fuwaṭī, Kamāl al-Dīn 'Abd al-Razzāḳ b. Aḥmad.," *Encyclopaedia of Islam, Second Edition* (Brill Online, 2010), http://www.brillonline.nl/subscriber/entry?entry=islam_SIM-3165.

[64] Muḥammad ibn Rāfi' Sallāmī, *Tārīkh 'ulamā' Baghdād al-musammá muntakhab al-mukhtār*, 2nd ed. (Bayrūt: Dār al-'Arabīyah lil-Mawsū'āt, 2000), 176; Aḥmad ibn 'Alī Ibn Ḥajar al-'Asqalānī, *al-Durar al-kāminah fī a'yān al-mi'ah al-thāminah*, 2nd ed. (al-Qāhirah: Dār al-Kutub al-Ḥadīthah, 1966), vols. 4, 59.

[65] Ibn Ḥajar al-'Asqalānī, *al-Durar al-kāminah fī a'yān al-mi'ah al-thāminah*, vols. 5, 108.; Franz Rosenthal, "Ibn Ḥadjar al- 'Asḳalānī, Shihāb al-Dīn Abu'l-Faḍl Aḥmad b. Nūr al-Dīn 'Alī b. Muḥammad.," *Encyclopaedia of Islam, Second Edition* (Brill Online, 2010), http://www.brillonline.nl/subscriber/entry?entry=islam_SIM-3178. Ibn Ḥajar al-'Asqalānī, *al-Durar al-kāminah fī a'yān al-mi'ah al-thāminah*, vol. 5; Rosenthal, "Ibn Ḥadjar al- 'Asḳalānī, Shihāb al-Dīn Abu'l-Faḍl Aḥmad b. Nūr al-Dīn 'Alī b. Muḥammad." in *Encyclopaedia of Islam, Second Edition (Brill Online, 2010)*, http://www.brillonline.nl/subscriber/entry?entry=islam_SIM-3178.

al-Dīn al-Subkī (d. 771 A.H.),[66] and *Ṭabaqāt al-shāfiʿīya*, by Jamal al-Dīn ʿAbd al-Rahīm al-Isnawī (d. 1332 C.E./772 A.H.).[67]

The biographical information on Shīrāzī found in Ibn al-Fuwaṭī, al-Sallāmī, and al-ʿAsqalānī also appears in later encyclopedic works such as *Kashf al-ẓunūn* by Kātip Çelebi (Hajjī Khalīfa, 1609–1657 C.E./1017–1067 A.H.),[68] *al-Badr al-Ṭāliʿ*, by al-Shawkānī (1760–1834 C.E./1173–1250 A.H.),[69] and *Rawḍāt al-jannāt*, by Muḥammad Bāqir Khwānsāsrī (1811–1895 C.E./1226–1313 A.H.).[70]

Historical Annals

Arranged chronologically, historical annals were records of the major political and social events of the year, and often included the passing away of significant individuals.[71] Annalistic works that mention Shīrāzī's name include: *al-Mukhtaṣar fī akhbār al-bashar*, by Ismāʿīl Abī al-Fidāʾ (1273–1331 C.E./672–732 A.H.),[72] *Mirʾāt al-janān wa-ʿibrat al-yaqẓān fī maʿrifat ḥawādith al-zamān*, by ʿAbd Allāh ibn Asʿad Yāfiʿī (c. 1298–1367 C.E./c. 698–768 A.H.),[73] and *al-Sulūk li maʿrifat*

---

[66] Tāj al-Dīn ʿAbd al-Wahhāb ibn ʿAlī Subkī, *Ṭabaqāt al-Shāfiʿīyah al-Kubrá*, 1st ed. (al-Qāhirah: al-Maṭbaʿah al-Ḥusaynīyah, 1324); J. Schact, "al- Subkī," *Encyclopaedia of Islam, Second Edition* (Brill Online, 2010), http://www.brillonline.nl/subscriber/entry?entry=islam_SIM-7116.

[67] ʿAbd al-Raḥīm ibn al-Ḥasan Isnawī, *Ṭabaqāt al-Shāfiʿīyah* (Baghdad: Riāsat diwān al-awqāf, 1971), 20.

[68] Kātip Çelebi, *Kashf al-ẓunūn ʿan asāmī al-kutub wa-al-funūn* (Bayrūt, Lubnān: Dār al-Kutub al-ʿIlmīyah, 1992); Orhan Şaik Gökyay, "Kātib Čelebi, appellation of muṣṭafā b. ʿabd allāh (1017–1067/1609–1657), known also (after his post in the bureaucracy) as ḥādjdjī khalīfa," *Encyclopaedia of Islam, Second Edition* (Brill Online, 2010), http://www.brillonline.nl/subscriber/entry?entry=islam_COM-0467.

[69] Muḥammad ibn ʿAlī Shawkānī, *al-Badr al-ṭāliʿ bi-maḥāsin man baʿda al-qarn al-sābiʿ* (Bayrūt, Lubnān: Dār al-Maʿrifah, 1978).

[70] Muḥammad Baqir Khvansari, *Rawdāt al-jannāt fī aḥwāl al-ʿulamā wa al-sadāt*, Rawdati, M., Ed. (Tehran: Dar al-Kutub al-Islamiyah, 1962); Abdul-Hadi Hairi, "Khwānsārī, Sayyid Mīrzā Muḥammad Bāḳir Mūsawī Čahārsūḳī b. mīrzā Zayn al-ʿābidīn," *Encyclopaedia of Islam, Second Edition* (Brill Online, 2010), http://www.brillonline.nl/subscriber/entry?entry=islam_SIM-4193.

[71] Franz Rosenthal, *A History of Muslim Historiography*, 2nd ed. (Leiden: E.J. Brill, 1968), 71.

[72] Abū al-Fidāʾ Ismāʿīl ibn ʿAlī, *al-Mukhtaṣar fī akhbār al-bashar*, 1st ed. (Baghdād: Maktabat al-Muthanná, 1968), pt. 4, 63. More typical however is the obituary note (*Ibid.*, part 4, p. 63; note that Shīrāzī's birthplace and the name of the *Tuḥfa* are rendered incorrectly): "On Sunday, the seventeenth of Ramadan the judge Quṭb al-Dīn Maḥmūd ibn Masʿūd [Shīrāzī] died in Tabrīz. And his birth was at Shayzar [sic] in the month of Ṣafar of the year 634, so he lived 76 years and 7 months and he was a renowned imām in a number of sciences such as mathematics and logic and the arts of medicine and principles of kalām and jurisprudence. He wrote a number of works including 'the limits of attainment in hay'a' and the 'Sāmī [sic] Offering' on hay'a, also, and ... his compositions and his virtues are well-known."

[73] ʿAbd Allāh ibn Asʿad Yāfiʿī, *Mirʾāt al-jinān wa-ʿibrat al-yaqẓān fī maʿrifat mā yuʿtabar min ḥawādith al-zamān*, 1st ed. (Bayrūt: Dār al-Kutub al-ʿIlmīyah, 1997), vols. 4, 187.

duwal al-mulūk, by Aḥmad ibn ʿAlī Maqrīzī (1364–1442 C.E./766–845 A.H.).[74] The entries in these works are generally short, giving the name, the occupation, and the date of Shīrāzī's death. However, the monumental *Tārīkh al-Islām*, by Shams al-Dīn Abū ʿAbd Allāh Muḥammad al-Dhahabi (1274–1348 C.E./673–748 A.H.) includes a substantial entry for Shīrāzī.[75]

Historical Chronicles on the Mongols in Iran and the Seljuks of Anatolia

Shīrāzī makes very few appearances in the sources that deal specifically with the history of the Mongols. These sources are important, however, in providing information on the social and political conditions of the world in which Shīrāzī lived. The standard works for the study of the Mongols in the Middle East are in Persian. The accounts of the campaigns of Chingiz Khan have been preserved in ʿAlāʾ al-Dīn ʿAtāʾ Malik b. Muḥammad Juwaynī's celebrated history, *Tarīkh-i jahān gushāy*, or *History of the World Conqueror*.[76] This book contains valuable information about the subsequent history of Persia up to the period immediately prior to the fall of Baghdād to the Mongols, in 1258 C.E.

Another centrally important work on the history of the Mongols for the period of interest is Rashīd al-Dīn Faḍl Allah's *Jāmiʿ al-Tawārīkh* or *Collection of Histories*.[77] This work is the primary historical source for the Ilkhanid dynasty as well as the events of Shīrāzī's era, and it was completed in 1310 C.E./710 A.H. A physician, and a convert from Judaism to Islam, this renowned *Ṣāḥib Dīwān* (or chief financial administrator) is also known as Rashīd al-Dīn Ṭabīb, in reference to his career as a physician prior to his entrance into the governmental bureaucracy. If the authenticity of a surviving collection of letters attributed to him is accepted it appears as though he, too, knew Shīrāzī personally.[78]

The geographer/historian Ḥamd Allāh Mustaufī Qazwīnī (d. after 1339–40 C.E./740 A.H.), who was a younger contemporary of Rashīd al-Dīn Ṭabīb, and who was appointed by him to work as a financial director in Qazwīn (modern Qazvin), also wrote a historical work encompassing the Ilkhanid period: the

[74]Aḥmad ibn ʿAlī Maqrīzī, *al-Sulūk li maʿrifat duwal al-mulūk*, 1st ed. (Bayrūt: Dār al-Kutub al-ʿIlmīyah, 1997), vols. 2, 164.

[75]Muḥammad ibn Aḥmad Dhahabī, *Tārīkh al-Islām wa-wafāyāt al-mashāhīr wa al-aʿlām*, U. Tadmurī, Ed. (Bayrūt: Dār al-Kitāb al-ʿArabī, 1987), vol. 54, 100.

[76]ʿAlāʾ al-Dīn ʿAṭā Malik Juwaynī, *The Ta'rikh-i Jahán-Gushá*, Qazvini, M., Ed. (Tehran: Bamdad, n.d.).

[77]Rashīd al-Dīn Ṭabīb, *Jāmiʿ al-tawārīkh*; Rashīd al-Dīn Ṭabīb, *Rashiduddin Fazlullah's Jamiʿu't-Tawarikh.*

[78]See Rashīd al-Dīn Ṭabīb, *Mukatabat-e Rashidi*, Panjab University oriental publications; (Lahore: Panjab University oriental publications, 1947).

Tārīkh-i Guzīda.[79] Qazwīnī completed this work in 1330 C.E./730 A.H. and dedicated it to one of Rashīd al-Dīn's sons.[80]

A list of historical works concerned with the Seljuks of Anatolia appears in The *Seljuks of Anatolia*, by Köprülü.[81] These works are also exclusively in Persian, and, with the notable exception of al-Ḥusayn b. Muḥammad al-Munshī' al-Ja'farī's *al-Awāmir al-'alā'īya* – commonly referred to as the *Saljūqnāmeh* – and the *Tadhkira-i Aqsārā'ī* (see below), have only been partly published.[82] al-Ja'farī is more well known by his pen-name Ibn Bībī and an abridged version of his history was published by Houtsma in 1902.[83] Houtsma's edition has also been reproduced in its entirety in *Akhbār-i salājeqe-i Rūm*, a compendium of Seljuk histories by Mashkur.[84]

Tadhkira-i Aqsārā'ī, the last section of a work entitled *Musāmarat al-akhbār* by Karīm al-Dīn Aqsārā'ī, contains a fair amount of information pertinent to Shīrāzī's life and career. It was completed in 1323 C.E./723 A.H. This work was published in 1983 in Tehran. (Portions of this work appear, as well, in Mashkur's work referred to previously.)

Cahen relies on these two works and others (including but not limited to a host of archival sources, as well as the aforementioned works by Juwaynī and Rashīd al-Dīn Ṭabīb) to compose the chapters in his *Pre-Ottoman Turkey* that are relevant to our study.[85] His historical narrative and the bibliography for the sections of his book dealing with the Seljuks and the Mongols are rich sources of information.

Mamluk Histories

The importance of Mamluk historians and their works for the understanding of the developments in Ilkhanid Persia has recently gained additional recognition.[86] This

---

[79]Ḥamd Allāh Mustaufī Qazvīnī, *Tārīkh-i Guzīdah*, Nawa'i, A., Ed. (Tehran: Amir Kabir, 1960), 701. The Quṭb al-Dīn Shīrāzī that is listed as having been executed by Ghāzān Khan in the year 700 A. H. (see p. 605) is clearly different from our Quṭb al-Dīn, highlighting one of the pitfalls of dealing with medieval histories such as we have listed.

[80]Farhad Daftary, *The Ismā'īlīs: Their History and Doctrines* (Cambridge [England]: Cambridge University Press, 1990), 330.

[81]Mehmet Fuat Köprülü, *The Seljuks of Anatolia: Their History and Culture According to Local Muslim Sources* (Salt Lake City, UT: University of Utah Press, 1992).

[82]Köprülü, *The Seljuks of Anatolia*, 10.

[83]Nāṣir al-Dīn Ḥusayn ibn Muḥammad Ibn Bībī, *Histoire des Seldjoucides d'Asie mineure d'après l'abrégé du Seldjouknāmeh d'Ibn-Bībī: Texte Persan, Publié d'après le Ms. de Paris* (Leiden: E.J. Brill, 1902).

[84]Nāṣir al-Dīn Ḥusayn ibn Muḥammad Ibn Bībī, *Akhbār-i Salājiqah-'i Rūm*, 1st ed. (Tehrān: Kitābfurūshī-i Tehrān, 1971).

[85]Claude Cahen, *Pre-Ottoman Turkey a General Survey of the Material and Spiritual Culture and History C. 1071–1330* (New York: Taplinger Pub. Co, 1968).

[86]David Morgan, "The Mongols in Iran: A Reappraisal," *Iran* 42 (2004): 131–136.

is especially true of episodes involving Ilkhanid–Mamluk relations. In 1282 Shīrāzī served as member of an embassy sent by the new Ilkhanid ruler Tegüder (Aḥmad Tekudār, in Persian sources) to Egypt. Substantial accounts of the arrival of this embassy and its reception at the court in Cairo appear in biographies of Sultan Qala'ūn by two Mamluk historians: *al-Faḍl al-ma'thūr min sīrat al-sulṭan al-malik al-manṣūr*, by Shāfi' ibn 'Alī ibn 'Asākir (1252–1330 C.E./649–730 A.H.),[87] and *Tashrīf al-ayyām wa al-'uṣūr fi sīrat al-malik al-manṣūr*, by Muḥyī al-Dīn ibn 'Abd al-Ẓāhir (1223–1292 C.E./620–692 A.H.).[88]

Other Histories

In his *Compendium of Dynastic Histories*, composed in Syriac and translated into Arabic with the title *Tārīkh Mukhtaṣar al-Duwal* Ibn al- 'ibrī (or Bar Hebraeus, 1225 or 1226–1286 C.E./623–685 A.H.) mentions Shīrāzī in a short list of luminary scientists of Ṭūsī's era. This confirms the claims by Shīrāzī's biographers as to his fame and renown during his own lifetime.[89] The interesting and in some cases unique accounts in Bar Hebraeus's history underline, as well, the importance of using non-Persian and non-Arabic sources for the study of Islamic history.

Shīrāzī's Works in Print

Three of Shīrāzī's books have been printed in recent times: two on philosophy, and one on medicine. The first is of the most relevance to our study. The other two are included here as corroboration of the earlier claim as to the unsatisfactory state of scholarship in regard to Shīrāzī. The texts are:

(a) The *Durra* (see section "Shīrāzī: preliminary remarks"): an encyclopedic philosophical work in Persian, dealing with logic, metaphysics, natural philosophy, mathematics, and theology.[90]
(b) *Sharḥ ḥikmat al-ishrāq* (or the "Commentary on the Philosophy of Illumination"), a commentary in Arabic on the great mystical philosopher Suhrawardī

[87] Shāfi' ibn 'Alī Ibn 'Asākir, *Šāfi' Ibn 'Alī's Biography of the Mamluk Sultan Qalāwūn*, Lewicka, P., Ed. (Warsaw: Dialog, 2000), 306–334.

[88] Muḥyī al-Dīn Ibn 'Abd al-Ẓāhir, *Tashrīf al-ayyām wa al-'uṣūr fi sīrat al-malik al-manṣūr*, 1st ed. (al-Qāhirah: Wizārat al-Thaqāfah wa al-irshād al-qawmī, al-idārah al-'āmmah lil-thaqāfah, 1961).

[89] Bar Hebraeus, *Tārīkh mukhtaṣar al-duwal* (Bayrūt: al-Maṭba'ah al-Kāthūlīkīyah lil-Ābā' al-Yasū'īyīn, 1890); Bar Hebraeus, *The Chronography of Gregory Abû'l Faraj, the Son of Aaron, the Hebrew Physician, Commonly Known as Bar Hebraeus*, Budge, E.A. Wallis, Ed. (London: Oxford Univ. Press, H. Milford, 1932).

[90] Shīrāzī, *Durrat al-tāj li-ghurrat al-dabāj*.

(1155–1191 C.E.). This is the best known commentary on Suhrawardī, and is the title most readily associated with Shīrāzī.[91]

(c) *Bayān al-ḥājah'ilā al-ṭibb wa al-aṭibbā' wa ādābuhum wa waṣāyāhum* (or the "Explication of the Need for Medicine and Physicians, Their Etiquette and Testaments") a short tract, a modern edition of which was published in Beirut in 2003.[92]

Secondary Sources from the Nineteenth and Twentieth Centuries

Shīrāzī is mentioned by a host of European historians of Islam and historians of science writing in the nineteenth and twentieth centuries. Information on him appears in Suter,[93] Wüstenfeld,[94] Brockelmann,[95] and Wiedemann.[96] He is also mentioned by Leclerc in an article that is based on the autobiographical note in Shīrāzī's commentary to Avicenna's *Canon*.[97]

Shīrāzī has also been the topic of a Ph.D. dissertation. Walbridge examined his commentary on Suhrawardī as a Ph.D. thesis written in 1983 at Harvard University.[98] Some of this material is re-examined in Waldridge's book *The Philosophy of Illumination*, published in 1992.[99] Material on Shīrāzī can also be found in biographies devoted to his illustrious teacher, Ṭūsī.[100] In addition, there are two extended biographies of Shīrāzī in Persian.[101]

---

[91]Quṭb al-Dīn Shīrāzī, *Sharh-i Hikmat al-Ishraq-i Suhravardi*, Nurani, 'Abd Allah., Silsilah-i danish-i Irani; 50; (Tihran: Muassasah-i Muṭala'at-i Islami, Danishgah-i Tihran, Danishgah-i Mak'gil, 2001).

[92]Quṭb al-Dīn Shīrāzī, *Bayan al-hajah ilá al-tibb wa al-atibba wa adabuhum wa wasayahum* (Bayrut: Dar al-Kutub al-'Ilmiyah, 2003).

[93]H. Suter, *Die Mathematiker und Astronomen Der Araber und Ihre Werke* (Leipzig: B.G. Teubner, 1900), 159.

[94]Ferdinand Wüstenfeld, *Geschichte Der Arabischen Aerzte und Naturforscher* (Göttingen: Vandenhoeck und Ruprecht, 1840), 148–149.

[95]Carl Brockelmann, *Geschichte Der Arabischen Litteratur Von Prof. Dr. C. Brockelmann* (Leiden: Brill, 1937), vols. 2, 510.

[96]E. Wiedemann, *Aufsätze zur Arabischen Wissenschaftasgeschichte* (Hildesheim: G. Olms, 1970).

[97]Lucien Leclerc, *Histoire de la Médecine Arabe* (Paris: E. Leroux, 1876), 129–130.

[98]John Walbridge, "The Philosophy of Qutb al-Din Shirazi; a study in the integration of Islamic philosophy." (Ph.D. diss., Harvard, 1983).

[99]Walbridge, *The Science of Mystic Lights*.

[100]See, for example, Mudarris Razavi, *Aḥwāl wa Athār-i Muḥammad Ibn Muḥammad Ibn al-Ḥasan al-Ṭūsī*, 136–141.

[101]Muhammad Taqi Mir, *Sharḥ-i ḥal wa āsār-i 'allamah Qutb al-Dīn Maḥmud Ibn Mas'ud Shīrāzi, danishmand-i 'ali qadr-i qarn-i haftum, (634–710 A.H.)*, Intisharat-i Danishgah-i Pahlavi 91; (Shiraz: Danishgah-i Pahlavi, 1977); M. nMinovi, "Mulla Qutb Shirazi," in *Yadnameh-i Irani-i Minorsky* (Tehran: Intisharat-i daneshgah-i Tehran, 1348), 165–205.

Chapter 2
The Mongols in Iran

Someone had fled Bukhārā after the event and came to Khurāsān. They asked him of the circumstances of Bukhārā. He said: "They came, they gouged, they burnt, they slew, they pillaged, and they left." The savvy crowd who heard this account agreed that greater concision could not be achieved in the Persian language.[1]

* * *

Chormaqan-qorchi subdued the Baqtat people. Knowing that the land was said to be good and its possessions fine, Ögödei-qahan issued the following decree: "Chormaqan-qorchi shall remain there as garrison commander. Each year he shall make [the people] send [me] yellow gold, [gold brocade], ... and damasks, small pearls, large pearls, sleek Arab horses...."[2]

* * *

And those who remained in the towns had for the most part blocked their doors with masonry, or partially barricaded themselves and entered and exited through the roofs, fleeing the tax-collectors. And when the tax-collectors would go to the neighborhoods they would reveal a miscreant low-life who had knowledge of the houses, and by whose guidance they could drag the people out of the nooks, cellars, orchards, and ruins.... As an example, the situation in Yazd was such that if one wandered its villages one could not see anyone at all to speak to or one from whom to ask directions. And the very few who had stayed behind had a designated lookout, who would signal as soon as he saw anyone at a distance, so that all could hide underground in the qanāts [i.e., aqueducts].[3]

---

[1]'Alā' al-Dīn 'Aṭā Malik Juwaynī, *Jahāngushā-yi Juvaynī: Changīz, Tārābī, Khvārazmshāh, Ḥasan Ṣabbāḥ, bā ma'nī-i vāzhah'hā*, 1st ed. (Tehrān: Intishārāt-i Mahtāb, 1371), 40; 'Alā' al-Dīn 'Aṭā Malik Juwaynī, *Genghis Khan: The History of the World Conqueror* (Seattle: University of Washington Press, 1997), 107.

[2]Urgunge Onon, *The Secret History of the Mongols: The Life and Times of Chinggis Khan* (Richmond, Surrey: Curzon, 2001), 267; Boyle, "Dynastic and Political History of the Īl-khāns," 107.

[3]Rashīd al-Dīn Ṭabīb, *Jāmi' al-tawārīkh*, 1028.

K. Niazi, *Quṭb al-Dīn Shīrāzī and the Configuration of the Heavens: A Comparison of Texts and Models*, Archimedes 35, DOI 10.1007/978-94-007-6999-1_2,

2.1 Introduction

The purpose of this chapter is to look at Mongol presence in Persia during the thirteenth century in order to better define the historical backdrop of Shīrāzī's life and career. While Shīrāzī was not yet born at the time of the initial conflict in the second and third decades of the century, the initial Mongol invasions were in many ways the defining events for the subsequent century and the trauma and disruption that they caused would likely have been felt not only by the immediate survivors but by subsequent generations, both in the affected areas and in neighboring regions. With the benefit of hindsight, historians often interpret the Mongol invasions and their aftermath as an attestation of the resilience of the subjugated cultures that were on the receiving end of the military campaigns of the Mongols. For the purpose of our study it is perhaps even more important to recognize that in this period the lives of those living in the eastern lands of the *abode of Islam*, whether cosmopolitan elites or illiterate peasants, abounded with various contingencies and uncertainties (as well, at times, as opportunities) that stemmed from their existence as imperial subjects of the vast Mongol empire. As a well-known scientist and scholar Shīrāzī spent much of his life close to the centers of political power and thus would have been particularly exposed to both the risks and rewards of the Mongol court.

Viewing the era through his lens of a world-historian living in the twentieth century, Marshal Hodgson terms the campaigns of Chingiz Khan and his successors the "Mongol Catastrophe." Yet, he concludes his discussion of the Mongol period on a positive note by emphasizing that, as traumatic as the Mongol invasions had been, their final result was the assimilation of the war-like nomads by the very cultures they had set out to conquer.[4] Other historians have noted as well the productive nature of the encounter between the Mongols and their Persian-speaking subjects, specifically with regard to the promotion of a pan-Asian trade network, the demand for luxury goods and the practice of relocating war prisoners (and the ensuing cultural cross-fertilization).

It is important, however, to not lose sight of what appears to have been the singularly violent nature of the initial conquests and the onerous political and economic conditions in the subsequent decades. The hindsight of our modern day observations with respect to the indefatigability of the beleaguered cultures of the eastern lands of Islam – their ability to grow, and to permeate neighboring regions, their success in attracting new religious adherents – should not cloud our perceptions, in other words, with respect to the cataclysmic nature of the period in question as they were perceived by those experiencing the Mongol campaigns and their aftermath.[5] Even though these campaigns created unprecedented opportunities

[4]Marshall G. S Hodgson, *The Venture of Islam: Conscience and History in a World Civilization* (Chicago: University of Chicago Press, 1974), vol. 2, 292.

[5]To gain some perspective on the situation in western Asia it should be noted that the campaigns of the Mongol armies in the first quarter of the thirteenth century appears to have resulted in the extermination of entire cultures, including that of the Tangut and Xi in Central Asia and China.

for the diffusion of goods and of ideas across Eurasia (considerable portions of which were to be ruled by a coalition of Mongol-ruled polities in the subsequent decades) and even though the rapid diffusion and close proximity of previously isolated cultures would no doubt have created a remarkable setting for cultural, religious and intellectual ferment, one of their most singular features remains their ferocity and violence, and – as far as Persia was concerned – the degree to which region was subjected (at least until the rule of Ghāzān, 1295–1304 C.E.) to a ruinous economic policy and exploitation.

Rather than do justice to the history of the Mongols in western Asia with its multiplicity of facets and profusion of detail (for which the reader is referred to the studies that appear in the bibliography) this chapter has the considerably more modest aim of presenting the major historical developments so as to provide a backdrop for our discussion of Shīrāzī's life. The primary goal remains, of course, to highlight especially those historical developments that would have been relevant to the life of Shīrāzī. Using the chronological scheme used by Boyle, we divide the period of interest into three phases: first, the period of the initial campaigns (1219–1226 C.E.); second, the period following the withdrawal of the main Mongol army with the installation of viceroys ruling in the name of the Great Khan in distant Mongolia (1226–1256 C.E., Boyle refers to this period as the period of the viceroys)[6]; third, the period of Ilkhanid rule in Persia (1256–1335 C.E.).[7] Though born during the period of the viceroys, Shīrāzī lived for essentially all of his adult life under Ilkhanid rule. Indeed, as we will see in Chap. 3 his association with Ṭūsī and Hülegü appears to have been shortly after the arrival of Hülegü in Persia, i.e., at the commencement of the third phase, as defined above. Yet, insofar as the claims to legitimacy by Hülegü and his successors were in many ways rooted in the conquests of Chingiz Khan, and the sociopolitical conditions of Persia had evolved out of those earlier episodes it is necessary to begin our discussion with the appearance of the Mongols in western Asia in 1219 C.E.

2.2 The Mongols in Iran: Global and Local Perspectives

Referring to the period from 945 to c. 1250 C.E. as the "Early Middle Era of Islamicate History," Hodgson characterizes it as one of prosperity and vigor.[8] He notes that many of the practices and institutions that are today associated with Islam were devised or, in having originated in the preceding period of the Abbasid "High Caliphate," came into their maturity during this period. As examples of such practices and institutions Hodgson lists the establishment of the '*ulamā*' as a social class, the spread of the Sufi orders, and the development of the *iqtā*' system of land

[6]Boyle, "Dynastic and Political History of the Īl-khāns," 106–109.

[7]Boyle, "Dynastic and Political History of the Īl-khāns," 106–109.

[8]Hodgson, *The Venture of Islam*, vol. 2, 4.

grants and of religious endowments or *awqāf*.[9] Having spent the previous period in a process of transformation, says Hodgson, the practices and institutions of the "Perso-Islamic" world coalesced into a normative form that was capable of being exported from its heartland, i.e., the land "between the Nile and the Oxus," to neighboring regions, e.g., Anatolia, North Africa, and across northern India, thus making this era one of expansion as well.[10]

Not surprisingly, if we were to examine the chronicles of a more local nature written by those who were living during Hodgson's Early Middle period, we would encounter periods that were less characterized by growth and prosperity than by reversal and discord. Indeed, in the strife-ridden accounts of the *fitna* (i.e., riots/discord) which led to the establishment of Seljuk power in Persia (c. 1040 C.E.) and the predations of the Turkish Ghuzz tribes in eastern Persia (c. 1150) one comes upon the record of appalling atrocities that resulted in widespread destruction.[11] The Ghuzz raiding campaigns in eastern Persia in 1179–1180 C.E., for example, are recorded in one of the local histories of Kirmān as follows:

> And when the Ghuzz succeeded in their designs, they surged out of Bāghayn and descended in the vicinity of the stream of Māhān, and when they had straitened the situation of Bardsīr [to its limit] they turned to Garmsīr and – Woe to the poor citizens of Jīruft, oblivious and unknowing! – for they swiftly descended upon them and annihilated one hundred thousand souls with a diversity of torments, trials, and tortures. Then, turning their attention to the countryside, wherever there was a prosperous region or an inhabited territory, they transformed it into denuded and abandoned rubble.[12]

Clearly, then, the difference in the two pictures, one depicting advance and the other recession is one of perspective: the first global and epochal, while the other local – both in the temporal and spatial senses.

That taken as a whole Hodgson's Early Middle period could be considered as a period of growth is especially remarkable, however, for the fact that this period was one in which the lands of Islam experienced a calamity that was of a bona fide global nature. This calamity, which was precipitated by the campaigns of the Mongol armies under their leader Chingiz Khan against their sedentary neighbors, started

[9]For a discussion of the *'ulamā* as a social class, see Hodgson, *The Venture of Islam*, vol. 2, 153; for the spread of sufi orders, see Hodgson, *The Venture of Islam*, vol. 2, 201; see also Hodgson, *The Venture of Islam*, vol. 2, 50–51.

[10]Hodgson, *The Venture of Islam*, vol. 2, 255–292.

[11]Omid Safi, *The Politics of Knowledge in Premodern Islam: Negotiating Ideology and Religious Inquiry* (Chapel Hill: University of North Carolina Press, 2006), 34; Hodgson, *The Venture of Islam*, vol. 2, 256; Bar Hebraeus, *The Chronography*, 202; 'Izz al-Dīn Ibn al-Athīr, *al-Kāmil fi al-tārīkh* (Báyrūt: Dar Ṣáder, 1385), XI, 176; Juwaynī, *Genghis Khan*, 285; Afḍal al-Din Kermani, *Badāyi' al-zamān fi waqāyi' Kirmān*, Bayani, M., Ed. (Tehran: Intisharat-i daneshgah-e Tehran, 1326), 88–89; Fakhr al-Dīn Gurgānī, *Masnavi-i Vis va Ramin* (Calcutta: College Press, 1865), 8–9.

[12]Kermani, *Badāyi' al-zamān fi waqāyi' Kirmān*, 89.

with attacks against the Chin dynasty, in northern China in 1213 C.E.[13] In western Asia the campaigns were slightly later with the attacks on the cities of Transoxiana commencing in 1219 C.E.[14] Though the parallels to the events surrounding the ascent of the Seljuks and the incursions of the Ghuzz tribes are readily apparent, the Mongol invasions (as recounted by the chroniclers of medieval Persia) dwarfed the scale of the earlier episodes in terms of severity as well as the geographical extent of the conflicts.[15] Indeed, even from a global perspective, these military campaigns appear to have been epoch-making, detrimentally affecting the prosperity of the subsequent two centuries – i.e., Hodgson's "Later Middle" period, 1250 to c. 1600 C.E.) across the entirety of the Eurasian continent.[16]

That the historical chronicles of the period are replete with accounts of extensive devastation or total destruction is an indication of the traumatic nature of these encounters in the shared experience of the chroniclers. What is particularly noteworthy in regard to the Persian historiography of the Mongols, however, is that in addition to references to "uncountable slayings"[17] and "the destruction of regions and the annihilation of the faithful"[18] one also encounters statements depicting devastation of such magnitude as to represent a woeful rupture with an irrecoverable past. Less than a century after the termination of Hodgson's Early Middle period, Mustaufī Qazwīnī writes: "There is no doubt that the destruction which happened on the emergence of the Mongol state and the general massacre that occurred at that time

[13]H. Desmond Martin, *The Rise of Chingis Khan and His Conquest of North China* (Baltimore: Johns Hopkins Press, 1950), 158. See Hugh Kennedy, *Mongols, Huns and Vikings: Nomads at War* (London: Cassell, 2002), 11, for a timetable of the Mongol conquests in China and Western Asia.

[14]Boyle, "Dynastic and Political History of the Īl-khāns," 307.

[15]*Encyclopaedic Ethnography of Middle-East and Central Asia*, 1st ed. (New Delhi: Global Vision Publishing House, 2005), 1; I. Petrushevsky, "The Socio-Economic Condition of Iran Under the Il-Khans," in *Cambridge History of Iran*, vol. 5, J. A. Boyle, Ed. (Cambridge: Cambridge University Press, 1968), 484; Hodgson, *The Venture of Islam*, vol. 2, 373.

[16]Noting a dearth of modern historical studies on the region, Hodgson is reluctant to blame the period of economic reversal in his "Later Middle period," i.e., subsequent to the Mongol campaigns, on a single cause. The discussion that appears under the rubric "the world-wide crisis" is suggestive but not conclusive: "For almost two centuries, there was something like a world depression reflected in the degree of urbanization, in the volume of trade, in the social resources available, even in sheer numbers of population. This may have been due partly to the after-effects of the Mongol devastations. These after-effects were both direct, in the lands that had themselves been devastated, and indirect, affecting the sources of world trade." Hodgson, *The Venture of Islam*, 2, 373. Hodgson adds: "The economy of the age of Mongol rule was not expansive but, at least in some areas, contracting – though (to what degree is not clear) on an Oikoumenic scale the Mongols themselves may have been partly responsible for this." Hodgson, *The Venture of Islam*, vol. 2, 386. The economy of the areas in Persia that were affected directly appeared to have suffered considerably, however. *See* I. P. Petrushevsky, "The Socio-Economic Condition of Iran Under the Il-Khans," in *Cambridge History of Iran*, vol. 5.

[17]Ḥāfiẓ Abrū, *Jughrāfiyā-yi Ḥāfiẓ Abrū: Qismat-i rub'-i Khurāsān, Harāt*, Māyil Haravī, R., Ed. (Tehrān: Bunyād-i Farhang-i Īrān, 1349), 33.

[18]Muḥammad ibn Aḥmad Nasawī, *Sirat Jalal al-Dīn Minkubirni*/Minuvi, Mujtabá, *Ed.* (Tihran: Shirkat-i Intisharat-i 'Ilmi va Farhangi, 1986), 79.

will not be repaired in a 1,000 years, even if no other calamity occurs; and the world will not return to the condition in which it was before that event."[19]

The most notable Persian work that chronicles this unprecedented set of encounters between the Mongols and the Persianate cultures of western Asia is the *History of the World-Conquerer* by ʻAlā' al-Dīn ʻAṭā Malik Juwaynī (1226–1283 C.E./623–681 A.H.).[20] Juwaynī commenced on writing this work c. 1252 C.E. and completed it c. 1260 C.E.[21] The book treats the history of the Mongols from shortly before Chingiz Khan's rise to power to the conquest of the Ismailis in Persia by Chingiz's grandson Hülegü. Juwaynī has been accused of servility to his Mongol patrons as well as of exaggerating the scale of the events he depicts. Though the accusations do not do justice to this remarkable historian and administrator,[22] there is no reason to doubt that Juwaynī would have had to accommodate both his urge to report the sensational and violent campaigns as well as his desire to please his patrons and to protect his own personal well-being, while cognizant at all times of his position as a high-ranking bureaucrat in the Mongol government. These facts may help explain why, for example, he is meticulous in recording the cities that were spared

[19]Ḥamd Allāh Mustaufī Qazwīnī, *The Geographical Part of the Nuzhat-al-qulub Composed by Hamd-Allāh Mustawfī of Qazwīn in 740 (1340)* (Leyden: E.J. Brill, 1915), 2, 34. The original Persian can be seen in the first volume of the same work: Qazwīnī, *The Geographical Part of the Nuzhat-al-qulub*, vol 1, 27.

[20]For Juwaynī's life see Barthold, W. "DJuwaynī, ʻAlā' al-Dīn ʻAṭā-Malik b. Muḥammad." *Encyclopaedia of Islam, Second Edition*. Edited by: P. Bearman; (Brill Online, 2011) http://www.brillonline.nl/subscriber/entry?entry=islam_SIM-2131. Bar Hebaraeus says of Juwaynī: "He had an adequate knowledge of the poetic art. And he composed a marvelous work in Persian on the chronology of the kingdoms of the Saljuks, and Khawarazmians, and Ishmaelites, and Mongols; what we have introduced into our work on these matters we have derived from his book." Bar Hebraeus, *The Chronography,* 473.

[21]George Lane, "JOVAYNI, ʻALĀ'-AL-Dīn," in *Encyclopaedia Iranica*, 2009, http://www.iranica.com/articles/jovayni-ala-al-Dn.

[22]See D. O. Morgan, "Persian Historians and the Mongols," in *Medieval Historical Writing in the Christian and Islamic Worlds*, D. O. Morgan, ed., (SOAS, London, 1982), 113–118. For the life of Juwaynī's first patron Möngke *see* Morgan, D.O. "Möngke." *Encyclopaedia of Islam*, Second Edition., Edited by: P. Bearman, (Brill Online, 2010) http://www.brillonline.nl/subscriber/entry?entry=islam_SIM-5260. In reading Juwaynī's history one can't help wondering if there aren't instances in which he may have reduced the level of mayhem and carnage a bit. The account of the wretched woman from Tirmidh, who, in an effort to buy time, admits to having swallowed some of her pearls, thus meeting an immediate and gruesome end, is one such example. The same account appears in Waṣṣāf's account. While it is true that Waṣṣāf's version is gorier and even more violent than Juwaynī, it is also more consistent with the level of mayhem in the rest of the account, and – given the tenor of the account – rings truer than Juwaynī's. It should also be noted that some of Juwaynī's astronomical figures may not have been too far off the mark. Jackson is one of the authors who disputes Juwaynī's figures for the number of descendants of Chingiz Khan, in his article "From Ulus to Khanate: The Making of the Mongol States c. 1220 – c. 1290," *The Mongol Empire and its Legacy*, Amitei-Preiss, R. and D. Morgan (Brill, Leiden, 1999), 12. Though Juwaynī's figures are implausibly high, modern genetic studies have in fact suggested a gargantuan number of offspring for the ruler (*see* Travis, J., "Genghis Khan's Legacy?," Science News 163, no. 6 (February 8, 2003): 91).

ruination.[23] That Juwaynī was interested, generally speaking, in the veracity of what he was relating can also be seen in the fact that occasionally – as in the episode of Khwārazm, he, too, encounters an unacceptably high figure for the dead and refuses to include it in his book.[24] So, while the purported scale of the destruction often seems implausible – at Marw (Merv) Juwaynī records 1,300,000 dead[25] – there is little reason to suspect Juwaynī of willfully inflating his figures. At any rate, to fully appreciate these figures it is important to recognize the true significance of the reports, i.e., that to witness as well as chronicler, the events precipitated by the Mongol invasions were of a singular and unprecedented scale, and the implausible figures that were reported by witnesses or chroniclers were meant to convey the unimaginable scale of the destruction.[26]

Juwaynī's loyalty to his employers as well as the recognition of his own place as a successful bureaucrat in the administration of the vast Mongol empire can perhaps best be discerned by the emphasis that he places on the efforts at rehabilitation since the original cataclysms (that had occurred roughly three decades before the time he was writing). This can be seen, for example, in his account of the sack of Bukhārā (1220 C.E.). Here Juwaynī provides a detailed account of the conquest of this important Central Asian city by describing the surrender of the townspeople, the resistance of the garrison stationed at the citadel, the use of the Bukhārans as human shields in the siege of the citadel, the filling of the moat (for the citadel) with the "animate and inanimate" bodies of the levied Bukhārans used as fodder, and the burning down of the entire town so that it came to resemble a "level plain."[27] Yet, he also concludes the same section of his work with a rather sanguine report of the subsequent revival of Bukhārā at the time of the penning of his book.[28]

There may be an additional significance to Juwaynī's buoyant tone in regard to the revival of Bukhārā, however, and this becomes apparent by reviewing his preliminary comments on the Mongol conquest of Transoxiana (in which both Bukhārā and Samarqand are located) as a whole:

> Chingiz Khan came to these countries in person. The tide of calamity was surging up from the Tartar army, but he had not yet soothed his breast with vengeance nor caused a river of blood to flow [as was pre-ordained by Fate]. When, therefore, he took Bukhārā and Samarqand, he contented himself with slaughtering and looting once only, and did not go to the extreme of a general massacre; and of those regions that were the dependencies and subsidiaries [i.e., of Bukhārā and Samarqand], since the majority of these offered their allegiance, [the Mongols] defiled these regions even less, and subsequently they

---

[23] Juwaynī, *Genghis Khan*, 89; Juwaynī, *The Ta'rikh-i Jahán-Gushá*, 69.

[24] Juwaynī, *Genghis Khan*, 128; Juwaynī, *Jahāngushā-yi Juvaynī*; Juwaynī, *The Ta'rikh-i Jahán-Gushá*, 101.

[25] Ibn al-Athīr's figure is 700,000, *al-Kāmil fī al-tārīkh*, 12, 393.

[26] D. Morgan, *Medieval Persia, 1040–1797* (London: Longman, 1988), 80.

[27] Juwaynī, *The Ta'rikh-i Jahán-Gushá*, 75–83.

[28] Juwaynī, *The Ta'rikh-i Jahán-Gushá*, 84–85.

> mollified what remained and were inclined to repair [these remains] so that presently [i.e., c. 1259/1260 C.E.] the prosperity and well-being of some of those domains equal what they were before, and for others they are approaching [their original condition].[29]

In a rather grim foreshadowing, however, Juwaynī continues:

> It is otherwise with Khurāsān and Iraq, which countries are afflicted with a hectic fever and a chronic ague: every town and every village has been several times subjected to pillage and massacre and has suffered this confusion for years, so that even though there be generation and increase until the Resurrection the population will not attain to a tenth part of what it was before. The history thereof may be ascertained from the records of ruins and midden-heaps declaring how Fate has painted her deeds upon palace walls."[30]

Here we see repeated (at a considerably smaller divide from the events themselves) a sense of the unspeakable horrors suffered by Khurāsān and Irāq (meaning here *'Irāq-i 'ajam*, or Persian "Iraq"),[31] and the enormous losses, economic as well as cultural, incurred by the lands that were on the Mongol war-path, as expressed by Qazwīnī.

It is reasonable to assume, then, that part of Juwaynī's project (his role as prominent bureaucrat notwithstanding) is to capture, within the general ghastliness of the war campaigns, a hierarchy of destruction and violence. Since by all accounts Khurāsān – the initial conquest of which Chingiz entrusted to his son, Tolui – appears to have borne the brunt of many of the exceptionally violent events during the conquest, Juwaynī may have been taking pains to make sure that the violence this region suffered was emphasized against the texture of the general mayhem.[32] In a short chapter entitled "A brief account of Toli's [i.e., Tolui's] Conquest of Khurāsān," Juwaynī writes:

> With one stroke a world which billowed with fertility was laid desolate, and the regions thereof became a desert and the greater part of the living dead, and their skin and bones crumbling dust; and the mighty were humbled and immersed in the calamities of perdition. And though there were a man free from preoccupations, who could devote his whole life to study and research and his whole attention to the recording of events, yet he could not in a long period of time acquit himself of the account of one single district nor commit the same to writing. How much more is this beyond the powers of the present writer who, despite his inclinations thereto, has not a single moment for study, save when in the course of distant journeyings, he snatches an hour or so when the caravan halts and writes down these histories![33]

---

[29]Juwaynī, *Genghis Khan*, 97 (slightly modified translation); Juwaynī, *The Ta'rikh-i Jahán-Gushá*, 75. Prof. Boyle's translation, with minor modifications.

[30]Juwaynī, *Genghis Khan*, 96; Juwaynī, *The Ta'rikh-i Jahán-Gushá*, 75. Prof. Boyle's translation.

[31]For a definition of *Irāq-i 'ajam*, see L. Lockhart, "DJibāl," in *Encyclopaedia of Islam*, Second Edition, Edited by: P. Bearman. (Brill Online, 2010), http://www.brillonline.nl/subscriber/entry?entry=islam_SIM-2068.

[32]Juwaynī, *The Ta'rikh-i Jahán-Gushá*, 144; al-Harawī Sayf ibn Muḥammad ibn Ya'qūb, *The Ta'ríkh Náma-I-Harat (The History of Harát) of Sayf Ibn Muḥammad Ibn Ya'qúb Al-Harawí* (Calcutta: Baptist Mission Press, 1944).

[33]Boyle, "Dynastic and Political History of the Īl-khāns," 734. Prof. Boyle's translation.

In contrast to Juwaynī's comments on the optimistic outcome at Bukhārā, this passage highlights the level of damage incurred by Khurāsān, while echoing as well Qazwīnī's sense of wonder and dismay at the magnitude of the destruction.

Chingiz Khan's final battle in western Asia, as it appears in Juwaynī's work, was against Sultan Jalāl al-Dīn, the last of the Khwārazmshāh dynasty – a Turkish dynasty ruling Persia – on the banks of the Indus (this is dated to between the 21st of August and the 19th of September, 1221 C.E.). This encounter was one from which Jalāl al-Dīn famously escaped with his life (eliciting, in so doing, the admiration and wonder of the Mongol ruler).[34] Until his death in August 1231 C.E., Jalāl al-Dīn represented the only tangible resistance to the Mongols, but this resistance – though perhaps significant to the immediate survivors of the Mongol campaigns – appears to have had little influence on the subsequent history of Persia.[35]

Shortly after his encounter with Jalāl al-Dīn, Chingiz turned his views homeward to distant Mongolia.[36] According to the *Secret History of the Mongols*, Chingiz "left governors at the cities he had conquered" before returning home.[37] Other historical sources state that in addition to the local governors (*basqāq* in Mongolian, *shaḥna/shiḥna* in Persian/Arabic) various Mongol generals acted as viceroys administering and conducting military operations within Persian lands in the period subsequent to Chingiz's return to Mongolia.[38] However, Judith Kolbas – whose research is focused on the numismatic evidence of the Mongol era – comments, on the absence of any evidence indicating a permanent Mongol presence south of the Oxus river, subsequent to the initial campaigns (i.e., Prof. Boyle's first phase). She suggests that the Mongol withdrawal, which may have in part been triggered by the Tangut uprising, changed at this point from a policy of "occupation" to "devastation." Returning to their Mongol homeland that had been made suddenly vulnerable by challenges and uprisings, Kolbas argues, the Mongol armies were left with no choice but to finish off any of the surviving populations that could provide resistance in the future.[39] If Kolbas is correct in her interpretation, then it is likely this scorched-earth policy with regard to the regions south of the Oxus river that is likely part of what survives in the chronicles as to the utter ruination of Khurāsān and *'Irāq-i 'ajam*.

Needless to say, the lack of a permanent Mongol presence in these regions would also help explain the accounts of the subsequent revival of Transoxiana, which as a permanent holding of the Mongols would likely have been subject to an official policy of repair and restoration. In this account, large portions of Persia to the south

---

[34]Boyle, "Dynastic and Political History of the Īl-khāns," 320.

[35]Boyle, "Dynastic and Political History of the Īl-khāns," 335.

[36]Boyle, "Dynastic and Political History of the Īl-khāns," 321.

[37]Onon, *The Secret History of the Mongols*, 254. See also the quote from *The Secret History* at the beginning of the current chapter.

[38]Boyle, "Dynastic and Political History of the Īl-khāns," 336–340.

[39]Judith G Kolbas, *The Mongols in Iran: Chingiz Khan to Uljaytu, 1220–1309* (London: Routledge, 2006), 60.

of the Oxus River – having been destroyed and heavily depopulated – may well have served primarily as a site for periodic looting raids or as grazing grounds for the large flocks of the pastoral Mongols. It is perhaps significant that in Rashīd al-Dīn's account, one of the only vassals listed as paying obeisance to Hülegü on the eve of his campaign in Persia were the Salghūrid ruler of Fārs, i.e., the region in which Shīrāzī was born and spent his youth and which appears to have been spared from destruction.

That during the era of the viceroys portions of Persia were left in a state of desolation with the absence of any semblance of a central authority can also be seen in local histories such as that of Ẓahīr al-Dīn Mar'ashī who writes of the northern region of Māzandarān: "And since the affairs of Māzandarān had remained in a state of lawlessness and chaos, Malik Ḥusām al-Daula ... conquered this region in the year 635 A.H. (1237–1238 C.E.), but since the [region] was, due to the decimation of the Mongols, empty of notables he was unable to provide order, and merely attempted to repair the cities and to provide law and order to the best of his ability. And he struck an agreement with the Rustamdār rulers to move to Amul, since the passage of the Mongol army was in Sārī."[40] Mar'ashī adds that the ruin heaps in Sārī and Amul were still visible when he was writing in 1470 C.E.[41]

It is also possible to discern from Mar'ashī 's words that, despite their vast scale, the Mongol campaigns in this period (i.e., during our first and second phases) were, characterized by some degree of unevenness with respect the degree of control exerted by the Mongols subsequent to their initial campaigns. As we noted earlier, Fārs which was Quṭb al-Dīn's birthplace, appears to have largely escaped destruction. Indeed, the Salghūrid rulers of Fārs appear to have been successful in negotiating a working relationship as vassals to the Mongols until the third quarter of the thirteenth century C.E.[42]

It is not clear to what extent the survivors of the Chingiz Khan's military campaigns (all of whom were now theoretically the subjects of the great Khan in distant Karakorum) could draw comfort from the fact that ruination had not visited all of the commercial and cultural centers of Persia to the same extent, and that the Ruler of the Faithful still ruled from Baghdād. At any rate, the political situation of the region was to change again with accession of Chingiz's grandson Möngke to the position of great Khan in 1251 C.E.[43] Seeking to consolidate the Mongol holdings in western Asia, he dispatched his brother Hülegü to the conquered lands in the west. Hülegü's campaign commenced in 1256 C.E. By 1258 the Ismaili polity in eastern and north-central Persia had been destroyed, Baghdād had been conquered and viciously sacked, the last caliph of the Abbasid line, executed. In addition all

[40]Ẓahīr al-Dīn Mar'ashī, *Geschichte von Tabaristan, Rujan und Masanderan* (St. Petersburg: Kaiserliche Akademie der Wissenschaften, 1850), 264.

[41]Ẓahīr al-Dīn Mar'ashī, *Geschichte von Tabaristan, Rujan und Masanderan*, 264.

[42]C. Bosworth, "Salghurids," in *Encyclopaedia of Islam*, Second Edition, Edited by: P. Bearman. (Brill Online, 2010), http://www.brillonline.nl/subscriber/entry?entry=islam_SIM-6531.

[43]Boyle, "Dynastic and Political History of the Īl-khāns," 340.

of modern-day Iran and much of present-day Iraq was incorporated into a newly formed Ilkhanid realm headed by Hülegü himself.[44] It is not apparent if the founding of the Ilkhanid polity was part of the original understanding with Möngke, but when this was accomplished it does not appear to have caused an issue with Karakorum.[45] Hülegü's descendants ruled Persia until their power disintegrated in the first half of the following century, nominally due to dynastic and succession issues, but no doubt, also due to practices and policies that ultimately proved unsustainable.

In the remainder of this chapter I will present a dynastic chronology of the Ilkhans, the dynasty under which – with the exception of the years of his youth – Shīrāzī was to spend all of his life and conclude with a review of the historical evidence of the observatory of Marāgha to discuss the role of the Ilkhans as patrons of the sciences and of astronomy in particular.

2.3 A Chronology of the Ilkhans

2.3.1 The Founding of a Dynasty: Hülegü (1256–1265 C.E.)

A grandson of Chingiz by Tolui, Hülegü[46] left Mongolia in 1253 at the behest of his brother the great Khan Möngke, with a mission to subjugate the Nizārī Ismaili's of Persia as well as subjugating the Abbasid caliph in the event that he refused to offer his allegiance.[47] Hülegü arrived at Samarqand in 1255 C.E., and received the homage of the minor rulers, amirs, and viceroys of Persia upon crossing the Oxus a short while later.[48] Among the rulers that paid homage were "the heir and successor of the Atabeg Muẓaffar al-Dīn of Fārs [i.e., the Salghūrid ruler], and the rival Seljuk sultans from Rūm, 'Izz al-Dīn and Rukn al-Dīn."[49] Hülegü's address to the assembly of amirs and atabegs appears in Rashīd al-Dīn's history:

> We have come to destroy the forts of the unbelievers by the Qā'ān's orders. If you have come of your will, with men and materiel, your land and home will remain yours, and your efforts will be appreciated. If not, then by God's will, when we are through with them we will march against you, heedless of excuses, and to your land and your home the same will be done as will have been done to theirs.[50]

---

[44]Boyle, "Dynastic and Political History of the Īl-khāns," 340–355.

[45]Boyle, "Dynastic and Political History of the Īl-khāns," 340.

[46]Hülegü is generally referred to as Hulākū or Hulāgū in the Persian sources, and as Hulāghū in Arabic sources.

[47]Rashīd al-Dīn Ṭabīb, *Jāmi' al-tawārīkh,* 688; R. Amitai, "HULĀGU KHAN," in *Encyclopaedia Iranica*, 2004, http://www.iranica.com/articles/hulagu-khan.

[48]Rashīd al-Dīn Ṭabīb, *Jāmi' al-tawārīkh*, 688.

[49]Boyle, "Dynastic and Political History of the Īl-khāns," 341; Rashīd al-Dīn Ṭabīb, *Jāmi' al-tawārīkh*, 688.

[50]Rashīd al-Dīn Ṭabīb, *Jāmi' al-tawārīkh*, 688.

The conquest of the Ismaili forts in Quhistān and Daylam, in eastern and north-central of Iran, proceeded swiftly and the Ismaili polity was effectively brought to an end with the surrender of the Ismaili ruler, Rukn al-Dīn Khūrshāh (r. 1255–1257 C.E.), at the fort of Maymūndiz on Sunday 29 Shawwāl 654 A.H./19 November 1256 C.E.[51] 'Alā' al-Dīn Juwaynī was present and, acting as Hülegü's secretary, penned the *yarligh* granting safe conduct to Rukn al-Dīn.[52] Upon the surrender of the fort of Alamūt some days later, Juwaynī was able to visit the famed library and to preserve some of the books and some of the astronomical instruments from destruction (while, at the same time, zealously consigning the Ismaili tracts that he found to the flames).[53] It was Juwaynī, also, who penned the terms of the surrender for the Ismailis.[54]

Naṣīr al-Dīn Ṭūsī was among the notables that surrendered at Maymūndiz.[55] The fame of this scientist, then in his fifties, had reached Karakorum, and Hülegü had been entrusted by the Great Khan with sending him to the Mongolian capital. Instead, Hülegü retained Ṭūsī as a member of his own retinue, where he became a trusted adviser and the administrator of the religious endowments (*awqāf*) in the Ilkhanid realms. Ṭūsī served as well as the first director of the Marāgha observatory; the construction of which was funded, at least according to some historians, by the very *awqāf* revenues for which Ṭūsī had been appointed as administrator.[56] In his *Zīj-ī Ilkhānī*, written during his tenure at Marāgha, Ṭūsī claims that he had been held by the Ismailis (whom he terms heretics) against his will, but this claim contradicts some of the other historical information from his life, including his own writings.[57]

Upon the extermination of the Ismailis Ṭūsī's new master, Hülegü, was able to focus on his second task: the extermination of the Abbasid caliphate. On the ninth of Rabī' al-ākhir 655 A.H. (April 25 1257 C.E.) he arrived at Dīnāwar and shortly thereafter at Hamadān where he sent a letter to the caliph "on the tenth of Ramaḍan, with warnings and promises (*bi tahdīd wa wa'īd*)," stating:

> At the time of the capturing of the forts of the infidels we asked for reinforcements from you; in response you claimed to be an ally, but did not send men.... Surely the word of men, common as well as exalted, has reached your ear as to what has befallen the world and its inhabitants at the hand of the Mongol armies from the time of Chingiz to the present time,

---

[51] Rashīd al-Dīn Ṭabīb, *Jāmi' al-tawārīkh*, 690; Juwaynī, *Genghis Khan*, 634; See also Daftary, *The Ismā'īlīs*, 426, and Kolbas, *The Mongols in Iran*, 155.

[52] Boyle, "Dynastic and Political History of the Īl-khāns," 344.

[53] Juwaynī, *Genghis Khan*, 719.

[54] Barthold, "DJuwaynī, 'Alā' al-Dīn 'Aṭā-Malik b. Muḥammad"; Rashīd al-Dīn Ṭabīb, *Jāmi' al-tawārīkh*, 697; Shafique N. Virani, "The Eagle Returns: Evidence of Continued Isma'ili Activity at Alamut and in the South Caspian Region Following the Mongol Conquests," *Journal of the American Oriental Society* 123, no. 2 (June 2003): 351–370.

[55] Rashīd al-Dīn Ṭabīb, *Jāmi' al-tawārīkh*, 695; Juwaynī, *Genghis Khan*, 635.

[56] Muḥammad ibn Shākir Kutubī, *Fawāt al-wafāyāt wa al-dhayl 'alayhā* (Beirut: Dar al-Thaqafah, 1973), 3, 250; Sayılı, *The Observatory in Islam and Its Place in the General History of the Observatory*, 207–211.

[57] Ṭūsī, *Naṣīr al-Dīn al-Ṭūsī's Memoir*, vol. 1, 10.

> and what humiliations were made to visit upon the Khwārazmshāhs, and Seljuks and the kings of Daylam and the Atabegs and others who were possessed of glory and might, at the hand of the eternal and ancient God. The gates of Baghdād were not secure against any of these factions, [so that] they held court there. Thus, given our might and power, how can they be secure against us?[58]

Given the fact that Rashīd al-Dīn lists concerns about both the (Ismaili) "unbelievers" as well as the "Caliph in Baghdād" as the reason for Hülegü's campaign, it is not clear if al-Musta'ṣam's cooperation would have changed the course of events.[59] At any rate, Baghdād fell to the Mongol army on the 4th of Ṣafar, 656 A.H. (February 10th, 1258 C.E.), signaling the end of the storied Abbasid dynasty that had served as the political and religious leadership of the Islamic *umma* for more than five centuries.[60]

Many secondary sources report that Hülegü chose Marāgha as his capital shortly after the fall of Baghdād.[61] The situation with primary sources is not as clear. Rashīd al-Dīn, the main authority on Hülegü's reign, mentions that Hülegü received the obeisance of vassals at Marāgha after the fall of Baghdād. However, neither Rashīd al-Dīn nor Waṣṣāf (another major source on Hülegü's reign) mention Marāgha as a

[58]Rashīd al-Dīn Ṭabīb, *Jāmi' al-tawārīkh*, 699.

[59]Rashīd al-Dīn Ṭabīb, *Jāmi' al-tawārīkh*, 684.

[60]Rashīd al-Dīn Ṭabīb, *Jāmi' al-tawārīkh*, 714. In regard to the extermination of the Abbasid line Rashīd al-Dīn states: "At the end of Wednesday on the fourteenth of Ṣafar of 656 they concluded the business of the caliph and his eldest son and five attendants who were with him, [at the village of *w-q-f*] and the following day, those of the others who had descended with him from the Kalwādhi gate, they martyred, and whomever of the Abassids they found, they did not leave alive, all except for the few whom they considered of no account. And Mubarakshāh the youngest son of the caliph they gave to Oljai Khatun, and Oljai Khatun sent him to Marāgha, to Khwājah Naṣīr al-Dīn, and they gave him a Mongol wife and he had two sons with her, and on Friday the sixth of Ṣafar they made the middle son of the caliph join his father and brother and the rule of the Abbasid caliphs who had come to power after the Umayyads was thus extinguished, and the period of their caliphate was five hundred and twenty five years." The caliph's death appears to have been in accordance with a Mongol practice that forbade the spilling of royal blood. This may be the source of the legend that the caliph died from hunger when he was imprisoned in a storeroom containing his treasure but no food. This account appears, for instance, in Waṣṣāf: 'Abd Allāh ibn Fazl Allāh Waṣṣāf-i Ḥazrat, *Geschichte Wassaf's* (Wien: Verlag der Österreichischen Akademie der Wissenschaften, 2010). A quote by the ruler of Mīyāfāraqayn alludes to this, and also to what must have been a perception that al-Musta'ṣam, had not allocated the proper funds for the defense of his domains: "Thanks be to God that I am not a dinar and dirham-worshipper like Musta'ṣam who lost his life and the kingdom of Baghdād due to his parsimony and miserliness." Rashīd al-Dīn Ṭabīb, *Jāmi' al-tawārīkh,* 725.

[61]C. Bosworth, "Ordu," in *Encyclopaedia of Islam*, Second Edition, Edited by: P. Bearman. (Brill Online, 2010), http://www.brillonline.nl/subscriber/entry?entry=islam_COM-0879; V. Minorsky, "Marāgha," in *Encyclopaedia of Islam*, Second Edition, Edited by: P. Bearman. (Brill Online, 2011), http://www.brillonline.nl/subscriber/entry?entry=islam_COM-0676; Rashīd al-Dīn Ṭabīb, *Jāmi' al-tawārīkh*, 714. Minorsky, V. "Marāgha." *Encyclopaedia of Islam*, Second Edition. Edited by: P. Bearman., (Brill Online, 2010) http://www.brillonline.nl/subscriber/entry?entry=islam_COM-0676.

capital city.[62] Indeed, Rashīd al-Dīn's chronicle suggests that Marāgha's privileged position may have been due in part to its selection by Ṭūsī as site of the observatory, the building of which commenced the same year as the fall of Baghdād:

> And in the aforementioned date, it was decreed, that the great Maulānā ... the sultan of the learned, Khwājah Naṣīr al-Dīn Ṭūsī (May the Lord conceal his faults through His mercy), in a location that he [saw] fit, set up a building for the observation of the stars. He chose a location in Marāgha.[63]

Certainly, little mention of this city is made in Rashīd al-Dīn's history in the subsequent accounts of Hülegü's life (which are primarily devoted to his various campaigns). These accounts describe Hülegü's attack on Syria,[64] his campaign against the Mamluks,[65] the campaigns of the Mongols in eastern Anatolia and the Caucasus,[66] the treachery of the son of Badr al-Dīn Lau' Lau' (the amir of Mosul) who allies himself with the Mamluks (and suffers a particularly gruesome death),[67] the outbreak of internecine warfare between Hülegü and Berke the Khan of the Golden Horde.[68] It is certain that for the majority of these episodes Hülegü would have been residing in his great mobile tent compound, or *ordū*.[69] Indeed, when Marāgha is mentioned again in the final chapter of Hülegü's life, it is in connection with the observatory (again suggesting that the observatory was what lent Marāgha its unique importance):

> Hulāgū loved buildings exceedingly, and of those that he has decreed many have survived. He built a palace in Alātāgh and built pagodas in Khoy and spent that year in the establishment of buildings and in the provident consideration of the welfare of the kingdom the army, and the populace. When fall arrived, desiring to establish his winter encampment at the Zarrīneh-rūd, [the river] which is called Jaghātū by the Mongols, he went to Marāgha and exerted his full efforts in the completion of the [observatory].[70]

---

[62]Rashīd al-Dīn's first mention of Marāgha, after the fall of Baghdād and the transfer of the loot from Baghdād, "and the forts of the unbelievers, and Rūm (Anatolia), and Georgia, and Armenia and the Lurs, and Kurds, likewise" to Azarbāijān, merely states that Hülegü received the obeisance of local rulers including Badr al-Dīn Lau' Lau' [the amir of Mosul] in the "vicinity of Marāgha." Rashīd al-Dīn continues "and sent him off on the sixth of Sha'bān of that year, and on the seventh ... the Atabeg Sa'ad the son of Abu Bakr the Atabeg of Fārs, offered his obeisance and felicitations on the conquest of Baghdād." See Rashīd al-Dīn Ṭabīb, *Jāmi' al-tawārīkh*, 717. However the two Seljukid amirs Izz al-Dīn and Rukn al-Dīn (who arrived subsequently) were received in a different locality (i.e., Mausaq, near Tabrīz). Rashīd al-Dīn Ṭabīb, *Jāmi' al-tawārīkh*, 717.

Tabrīz appears to have become the official capital of the Ilkhanid dynasty under Hülegü's successor, Abaqa, shortly after his accession on June 19, 1265/third of Ramadan 663. *Jāmi' al-tawārīkh*, 742–743.

[63]Rashīd al-Dīn Ṭabīb, *Jāmi' al-tawārīkh*, 718.

[64]Rashīd al-Dīn Ṭabīb, *Jāmi' al-tawārīkh*, 719–725.

[65]Rashīd al-Dīn Ṭabīb, *Jāmi' al-tawārīkh*, 721–725.

[66]Rashīd al-Dīn Ṭabīb, *Jāmi' al-tawārīkh*, 725–729.

[67]Rashīd al-Dīn Ṭabīb, *Jāmi' al-tawārīkh*, 729–731.

[68]Rashīd al-Dīn Ṭabīb, *Jāmi' al-tawārīkh*, 731–734.

[69]Linda Komaroff, ed., *Beyond the Legacy of Genghis Khan* (Leiden: Brill, 2006), 5.

[70]Rashīd al-Dīn Ṭabīb, *Jāmi' al-tawārīkh*, 734.

According to Rashīd al-Dīn, of the amirs that Hülegü received in Marāgha after the fall of Baghdād were the governors of Shīrāzī's home province of Fārs (the Atabeg Sa'ad) as well as the brothers Rukn al-Dīn and 'Izz al-Dīn, who were rival Sultans in Rūm (Anatolia) having been installed in 1246 C.E.[71]

Subsequent to the sack of Aleppo and Damascus by the Mongols in 1259, news of the death of Möngke caused Hülegü to withdraw a portion of his forces to the east. Subsequently his general Kitbughā was defeated by the Mamluks of Egypt at 'Ayn Jalūt (i.e., "the spring of Goliath").[72] This was a significant reversal of Ilkhanid fortune, for it halted the westward advance of the Mongol military machine, and established the Euphrates as the boundary between the two polities. It confirmed as well, the Mamluks as the primary rival for Mongol hegemony in the eastern Mediterranean – a rivalry that was to last for the remainder of the Ilkhanid era.[73]

The Mamluk Turks – themselves of a Central Asian and nomadic background – had begun to consolidate their power upon the appointment of one of their members, Qutuz, to the regency of Egypt in the aftermath of the defeat of the French monarch Louis IX and his fellow crusaders.[74] Mamluk-Mongol relations were to greatly preoccupy the subsequent Ilkhan rulers; at least until Öljeitü's last campaign against them in 1313 C.E.[75] These relations were bitterly antagonistic, and were the cause of repeated attempts by the Mongols and European armies, both within the crusader states in Syria and in Europe proper, to form alliances with each other, against the Mamluks.[76] The Mongol defeat at 'Ayn Jalūt, which had followed a less crushing defeat of a smaller Mongol force in Gaza (where the Mamluks had again been led by Qutuz), was followed by yet another defeat on the 10th of December 1260 C.E., at Ḥimṣ. Baybars, who had led the Mamluk army to victory at Ḥimṣ, and who had been instrumental in the victory at 'Ayn Jalūt had by then become the new Mamluk ruler; having assassinated Qutuz in the short interval between 'Ayn Jalūt and Ḥimṣ.[77] He was to be an indefatigable opponent of the Mongols until his death in 1277 C.E.[78]

---

[71]Cahen, *Pre-Ottoman Turkey*, 271–273. These figures are the very same who greeted Hülegü on his arrival (see note 48), suggesting a possible duplicate rendition of the same event.

[72]Boyle, "Dynastic and Political History of the Īl-khāns," 352.

[73]Boyle, "Dynastic and Political History of the Īl-khāns," 352.

[74]Syedah Fatima Sadeque, *Baybars I of Egypt*, (Dacca, Oxford University Press, 1956), 36.

[75]Boyle, "Dynastic and Political History of the Īl-khāns," 403.

[76]Constantinople was reclaimed by the Byzantines from the Latins in 1261, leaving the crusader cities of the Levant as the only crusader presence in the eastern Mediterranean. See R.L. Wolff, "The Latin Empire of Constantinople, 1204–1261," in *The History of the Crusades*, vol. 2 (Philadelphia: University of Pennsylvania, 1962), 231–233.

[77]Sadeque, *Baybars I of Egypt*, 39–42. For the origin of the term *al-Bunduqdāri* or *Bunduqdār*, the title by which Baybars was known (and by which Rashīd al-Dīn refers to this energetic and successful ruler) see Sadeque, *Baybars I of Egypt*, 30. Homs is one of two English spellings for this important Syrian city, which is also commonly referred to as Hims.

[78]Sadeque, *Baybars I of Egypt*, 46–54, 64–69.

In the last chapter on Hülegü's life Rashīd al-Dīn describes the manner in which he delegated the rule of his vast conquests, consigning Iraq, Khurāsān, and Māzandarān to the shores of the Oxus to his "oldest and best son," Abaqa, and "Arrān and Azarbāijān ... to Prince Yashmūt, and Diyārbakir and the Rabī'a region up to the Euphrates to the Amir Tudān, and Rūm to Mu'īn al-Dīn Suleimān Parvāneh."[79] As we will see, Mu'īn al-Dīn was to become one of Shīrāzī's patrons. In Chap. 3 we use the date of Mu'īn al-Dīn's execution (in 1277 C.E., by the order of Hülegü's son, Abaqa) to help pin some of the dates in Shīrāzī's life. Rashīd al-Dīn states that Hülegü assigned Shīrāzī's home region of Fārs – ruled as we saw by the Salghūrid dynasty who were vassals to the Mongols – to the Amir Iknānū, presumably as an overseer of Mongol interests in that vassal state.[80] Hülegü selected Shams al-Dīn Muḥammad Juwaynī, Ata' Malik Juwaynī's brother (and subsequently a patron of Shīrāzī), as the vizier of his domains, "granting him full and absolute power in the [administration of the] kingdom."[81] The author of the *History of the World Conquerer* himself was granted the important governorship of Baghdād.[82]

Hülegü's death occurred in the year 663 A.H. (1265 C.E.):

> As the year of the Bull arrived in the Rabī' al-Awwal of the year 663 (Dec. 1264/Jan. 1265) he was busy with hunting and festivities (*tūy*). Suddenly after the bath an illness returned to his body, through which he felt heavy and became bedridden. And on Tuesday the seventh of Rabī' al-ākhir he took from the hand of the Chinese doctors a laxative, which resulted in unconsciousness and led to a stroke. And no matter how diligently the capable doctors did attempt the purge they were unable to deflect the malady since the levels of vitality had reached the point of morbidity, and no fateful arrangement could be found that was fruitful, nor could a providential drug be found that was beneficial. And at that time a comet came into view, shaped as a conical rod, appearing every night, and as it disappeared on Sunday night of the nineteenth of Rabī' al-ākhir of the year 663 [A.H.] the great event took place. His age was 48 full solar years and on the banks of the Jaghātū he left the roadhouse of annihilation for the eternal abode.[83]

Hülegü's funeral appears to have been the last Mongol burial in Persia involving human sacrifice. Rashīd al-Dīn discretely omits any mention of this, simply stating: "They built his tomb in the Shāhī mountain that faces Dehkhārghān and in his camp they held mourning ceremonies, and buried his coffin in the tomb."[84] The reference

---

[79] Rashīd al-Dīn Ṭabīb, *Jāmi' al-tawārīkh*, 734.

[80] Rashīd al-Dīn Ṭabīb, *Jāmi' al-tawārīkh*, 734; Thackston's rendition of this name is Vangianu. See Rashīd al-Dīn Ṭabīb, *Rashiduddin Fazlullah's Jami'u't-Tawarikh*, 2, 513.

[81] This he does after executing Amir Sayf al-Dīn Batikchī, the previous holder of the post. See Rashīd al-Dīn Ṭabīb, *Jāmi' al-tawārīkh*, ibid. We can only speculate on how the administrative duties of Shams al-Dīn may have affected Ṭūsī's role as chief administrator of the religious endowments. Certainly that Shams al-Dīn's brother does not mention Ṭūsī in his accounts of the fall of the Ismailis is one of the striking omissions in the *World Conqueror*.

[82] Rashīd al-Dīn Ṭabīb, *Jāmi' al-tawārīkh*, 734.

[83] Rashīd al-Dīn Ṭabīb, *Jāmi' al-tawārīkh*, 736.

[84] Rashīd al-Dīn Ṭabīb, *Jāmi' al-tawārīkh*, 736.

appears rather in Waṣṣāf: "And in the manner of the Mongols they built a crypt, and poured great quantities of jewels and gold in it, and several ravishing beauties were made to accompany him in his eternal sleep, so that he would be immune to the fear of oblivion."[85]

2.3.2 *The Mamluk Challenge: Abaqa (1265–1282 C.E.)*

The day for Abaqa's accession ceremony was determined by Khwājah Naṣīr al-Dīn to be the third of Ramadan, 663 A.H. (June 19th, 1265 C.E.) with Virgo ascendant (*bi ṭali'-i sunbula*).[86] Despite the purported auspiciousness of this day, Abaqa was soon faced with threats from the neighboring Mongol factions of the Golden Horde, and the Chaghatai Khanate of Central Asia.[87] The conflict with the Golden Horde was resolved in 1266 C.E. with the death of Berke, Abaqa's uncle and the Khan of the Golden Horde.[88] The Chaghatai armies were dealt a bloody defeat at Harāt on the first of Dhū al-Ḥijja 668/22 July 1270; though raiding parties from Central Asia continued to menace the eastern regions of the Ilkhanate periodically.[89]

Abaqa appears to have taken over the rulership of the Ilkhans with the unanimous support of the Ilkhanid nobles, yet had to wait for confirmation by the great Khan, Qubilai who had succeeded his brother Möngke and had consolidated his rule against the majority of his rivals by 1264 C.E.[90] Rashīd al-Dīn states that "despite being the protector (*walī*) [i.e., the rightful owner] of the crown and the throne – until the arrival of the messengers from his highness Qubilai Khan and their bringing the *yarligh* in his name – he conducted his affairs seated on a chair." The *yarligh* with Qubilai's endorsement did not arrive until 1270.[91] This may explain why, upon his (first, unofficial) accession, Abaqa was munificent to the extreme. According to Rashīd al-Dīn "he gave an untold amount of money and jewelry and fine clothing to the courtiers (*khawātīn*), the princes and the amirs, so much so that [even] most of the soldiery was able to benefit."[92] In addition, "he made nearly one

---

[85]'Abd Allāh ibn Fazl Allāh Waṣṣāf al-Ḥazrat, *Geschichte Wassaf's* (Wien: Verlag der Österreichischen Akademie der Wissenschaften, 2010), 101.

[86]Rashīd al-Dīn Ṭabīb, *Jāmi' al-tawārīkh*, 742; Kutubī, Fawāt al-wafāyāt wa al-dhayl 'alayhā, 3, 249.

[87]P. Jackson, "ABAQA," in *Encyclopaedia Iranica*, *Encyclopaedia Iranica Online*, 1982, http://www.iranica.com/articles/abaqa.

[88]Peter Jackson, "ABAQA."

[89]Peter Jackson, "ABAQA."

[90]Barthold, W. "Ḳubilay." *Encyclopaedia of Islam*, Second Edition. Edited by: P. Bearman, (Brill Online, 2010) http://www.brillonline.nl/subscriber/entry?entry=islam_SIM-4469.

[91]Rashīd al-Dīn Ṭabīb, *Jāmi' al-tawārīkh*, 765. Also see Peter Jackson, "ABAQA," in *Encyclopaedia Iranica.*

[92]Rashīd al-Dīn Ṭabīb, *Jāmi' al-tawārīkh*, 743.

hundred well-known scientists who were the students of the teacher of mankind, Khwājah Naṣīr al-Dīn Ṭūsī, May the Lord have mercy upon him, the beneficiaries of an all-embracing boon."[93]

Despite threats by his kinsmen in the Caucasus and Central Asia, the adversaries that were to demand the most attention during Abaqa's rule were the Mamluks. The intense rivalry of these two polities played itself out repeatedly in Syria and in Anatolia throughout Abaqa's reign. The first twelve years of Abaqa's reign coincided with the reign of Baybars. By 1261 C.E. Baybars had re-established Mamluk control over Damascus and Aleppo, and had had a new caliph installed in Cairo to help legitimize his rule.[94] He had also formed an alliance with Berke, the Khan of the Golden Horde, in 1264 C.E. In 1267 C.E. a skirmish with the Mongols under their new ruler Abaqa ended in a retreat of the Mongol forces.[95] In the face of such an energetic adversary, Abaqa in turn sought an alliance with Prince Edward of England (later King Edward I) who was leading the crusaders against the Mamluk armies. This alliance was not particularly fruitful, however, since the size of the Mongol forces that were dispatched was apparently too small.[96] In 1277 Baybars invaded Rūm, roundly defeating the Mongol army at Abulustān.[97] In retaliation for the tepid support of his Seljuk vassals Abaqa ordered the destruction of the area between Qaisarīya and Erzurum, in the same year; calling off the slaughter and the mayhem only after the *Ṣaḥib Dīwān* Shams al-Dīn's intervention.[98] Mu'īn al-Dīn Suleimān (also known as the keeper of the seals or "the Parvāneh"), whom as we saw had been confirmed in his role the Mongol-appointed administrator of Rūm by Hülegü, was accused of supporting the Mamluk attack, and paid with his life for this alleged intrigue with Baybars.[99]

Rashīd al-Dīn states that in addition to leaving Shams al-Dīn in power as the chief administrator of the Mongol realms at the beginning of his reign, Abaqa appointed his son, Bahā' al-Dīn Muḥammad as the governor of *'Iraq-i 'ajam*.[100] Bahā' al-Dīn continued his service under Abaqa, until his death in the year 678 A.H. (1279/1280 C.E.).[101] In his introduction to the *Durra* Mishkat identifies Bahā' al-Dīn as the dedicatee of Shīrāzī's *Nihāya*.[102] This identification creates an immediate chronological problem and (if the date of Baha' al-Dīn's death is accepted as valid)

[93]Rashīd al-Dīn Ṭabīb, *Jāmi' al-tawārīkh*, 744.

[94]Sadeque, *Baybars I of Egypt*, 43–46.

[95]Sadeque, *Baybars I of Egypt*, 57.

[96]Jackson, "ABAQA"; Michael Prestwich, *Edward I* (New Haven: Yale University Press, 1997), 78.

[97]Boyle, "Dynastic and Political History of the Īl-khāns," 361.

[98]Boyle, "Dynastic and Political History of the Īl-khāns," 361.

[99]Boyle, "Dynastic and Political History of the Īl-khāns," 361.

[100]Rashīd al-Dīn Ṭabīb, *Jāmi' al-tawārīkh*, 744.

[101]Juwaynī, *The Ta'rikh-i Jahán-Gushá*; 'Abd Allāh ibn Fazl Allāh Waṣṣāf al-Ḥaẓrat, *Taḥrīr-i tārīkh-i Waṣṣāf* (Tehran: Bunyād-i Farhang-i Īrān, 1967), 34–37.

[102]Quṭb al-Dīn Shīrāzī, *Durrat al-tāj li-ghurrat al-dabāj*. page n.

cannot be correct.[103] (We will revisit the problem of identifying the dedicatee of the *Nihāya* in Chap. 3.) Shams al-Dīn and his brother had to contend with forceful attempts by fellow courtiers to dislodge them from their positions of prominence. In addition to being charged with embezzlement, the brothers were charged with the perhaps even more serious crime of harboring pro-Mamluk sympathies. ʻAlā' al-Dīn was punished by being humiliatingly paraded in Baghdād, and was subsequently imprisoned in Hamadan.[104] Indeed Abaqa's death in Hamadān on the twentieth of Dhū al-ḥijja 680/April 1st 1282 C.E., after an evening of excessive drinking, would no doubt have come as a welcome reprieve for both Juwaynī brothers.[105]

2.3.3 An Adoption of Popular Customs: Tegüder Aḥmad (1282–1284 C.E.)

A notable feature of the reign of Tegüder Aḥmad (or Takūdār, in the Persian sources) is his conversion to Islam (whence the Arabic name Aḥmad), is reported rather tepidly in the account by the Syrian historian Abū al-Fidā'. "And when Abaqa died, his brother Aḥmad the son of Hülegü became king and the name of this aforementioned Aḥmad was Biker [sic], and since when he assumed power he professed Islam he was called Aḥmad Sultan."[106] As a Mamluk official the lukewarm tone in Abū al-Fidā''s report is perhaps understandable. Rashīd al-Dīn appears to be as unimpressed as Abū al-Fidā', however: "They sat him on the throne, and celebrated in the manner to which the Mongols are accustomed, and since he professed Islam they called him Sultan Aḥmad."[107] This presentation is in stark contrast with that of Rashīd al-Dīn's employer Sultan Ghāzān, whose conversion to Islam is praised by Rashīd al-Dīn with a lofty and ornate language. One of the possible reasons for the ambivalence regarding Aḥmad's profession of Islam is the questionable reputation of the man said to be responsible for his conversion: Tegüder Aḥmad's "adviser," Sheikh ʻAbd al-Raḥmān of Mosul, was considered by

---

[103] As we will see the *Nihāya* was completed in November of 1281 C.E. and so postdates Bahā' al-Dīn's death by approximately a year.

[104] Rashīd al-Dīn Ṭabīb, *Jāmiʻ al-tawārīkh*, 774.

[105] Biran, "JOVAYNI, ṢĀḤEB DIVĀN," in *Encyclopaedia Iranica*, 2009, http://www.iranica.com/articles/jovayni-saheb-divan; Rashīd al-Dīn Ṭabīb, *Jāmiʻ al-tawārīkh*, 779; Jackson, "AḤMAD TAKŪDĀR," in *Encyclopaedia Iranica*, Encyclopaedia Iranica Online., 1984, http://www.iranica.com/articles/ahmad-takudar-third-il-khan-of-iran-r.

[106] Ismāʻīl ibn ʻAlī Abū al-Fidā', *al-Mukhtaṣar fī akhbār al-bashar*, pt. 4, 63 (See Chap. 1, note 72).

[107] Rashīd al-Dīn Ṭabīb, *Jāmiʻ al-tawārīkh*, 785. It is understandable that Rashīd al-Dīn saves his most fulsome accolades for the conversion of his own employer, Sultan Ghāzān. In addition the copier of the manuscript available for the Karimi edition appears to have had a personal experience with Sultan Aḥmad; a petition of his for which he nearly pays with his life, and includes an account of this encounter as a reprobation of Aḥmad. See Rashīd al-Dīn Ṭabīb, *Jāmiʻ al-tawārīkh*, 801.

many to be a charlatan.[108] In Rashīd al-Dīn's description, the Sheikh is depicted as something of a distraction to Aḥmad's official duties.

> [Aḥmad] had a great intimacy with 'Abd al-Raḥmān, so much so that he called him bābā [i.e., father], and he called Ishan Manklī who was a follower of Bābī Ya'qūb, who had a station in Arrān, qarandash, and would go to their house at all times (Ishan Manklī's house was in the back of the [Royal encampment]) and participate in the *simā'*. And he was less likely to attend to the organization and arrangement of governmental issues, and his mother Qūtī Khātūn who was wise and capable to the extreme, ensured the interests of the various realms were met.[109]

The Sheikh is important for our study, since Rashīd al-Dīn states that "it was at the suggestion of the Sheikh 'Abd al-Raḥmān and Shams al-Dīn (i.e., Juwaynī) the *Ṣāḥib Dīwān*, that [Sultan Tegüder Aḥmad] sent Maulānā Quṭb al-Dīn Shīrāzī who was a learned man as a messenger to Egypt on the nineteenth of Jumāda I, 681 (Aug. 25, 1282 A.H.)."[110] This embassy, which undoubtedly signifies the prestige of Shīrāzī as a scholar in the court of Tegüder, was the first of two sent by Tegüder Aḥmad. The embassy conveyed a written message which appears in full in Shāfi's account (and is described by him as clattering "with the clatter of the *'ajam*").[111] It opens with thanks to the Lord for guiding the ruler to Islam, and describes Tegüder's desire for peace – despite a Mongol assembly (Kuriltai) in which the notables had voiced their desire for a continuation of Abaqa's antagonism with the Mamluks. It lists, as well, Tegüder's reforms which had allowed for improvements in providing for the welfare of his subjects.[112] Modern historians have generally viewed the

---

[108]R. Amitai, "Sufis and Shamans: Some Remarks on the Islamization of the Mongols in the Ilkhanate," *Journal of the Economic and Social History of the Orient* 42, no. 1 (1999): 27–46; Shāfi' ibn 'Alī Ibn 'Asākir, *Šāfi' Ibn 'Alī's Biography of the Mamluk Sultan Qalāwūn*, 308.

[109]Rashīd al-Dīn Ṭabīb, *Jāmi' al-tawārīkh*, 788. Sheikh 'Abd al-Raḥmān is also described as a person with supernatural powers. In an episode depicting the intrigue of the courtier Majd al-Mulk against his patrons the Juwaynī brothers we read: "A decree was passed stipulating the return [to their owner] of the possessions and articles of Khwājah 'Ala' al-Dīn Ata' Malik [Juwaynī] that had been ... confiscated .... 'Ala' al-Dīn prepared them and presented them [stating]: "What we brothers have accomplished has been through the all-encompassing blessing of the Ilkhāns. In this quriltai [i.e., assembly] your servant [willingly disburses these items back to the treasury]".... And it was decreed that Majd al-Mulk [stand trial instead] .... [During the trial] in the midst of his trappings they found a fragment of a lion's skin, upon which something had been written in yellow and red with an illegible hand, and since the Mongols detest sorcery to the extreme, they were terrified of the script .... The ... sorcerers said that the protective charm should be doused with water, and that [Majd al-Mulk] be forced to drink the extract so that the magical evil would be neutralized. And they prompted Majd al-Mulk to carry this out, but he refused, since the protective charm was one that Sheikh Abd al-Raḥman had devised, and [one he] had planted in his trappings and he was sure that it could not be devoid of [evil powers]." Rashīd al-Dīn Ṭabīb, *Jāmi' al-tawārīkh*, 787. See also Bar Hebraeus, *The Chronography*, 474; and Amitai, "Sufis and Shamans."

[110]Rashīd al-Dīn Ṭabīb, *Jāmi' al-tawārīkh*, 787.

[111]Shāfi' ibn 'Alī Ibn 'Asākir, *Šāfi' ibn 'Alī's Biography of the Mamluk Sultan Qalāwūn*, 309. This letter was likely written by Shīrāzī himself as we will see in Chap. 3, Sect. 3.5.

[112]Shāfi' ibn 'Alī Ibn 'Asākir, Šāfi' ibn 'Alī's Biography of the Mamluk Sultan Qalāwūn, 309–316.

embassy as a gesture of peace by the newly converted Mongol ruler.[113] However, the presence of a fragment of verse 17:15 of the Qur'an, "And we do not mete out torment until after we have sent a messenger [to warn]" in the closing of the letter, as well as other features have led one modern historian to conclude that the letter is actually a sort of ultimatum by the Mongol khan to the Mamluk ruler.[114] In any event, the mission was a failure, either as ultimatum or indeed as far as changing the status quo between the warring states.

The Mamluk historian 'Abd al-Ẓāhir writes of the embassy that it was a large one, consisting of "subjects, escorts, slave boys, slave soldiers and notables, all in great splendor."[115] He adds: "When they had reached Bira [on the Euphrates, i.e., the frontier] the Sultan wrote to his deputies to guard against them and [to ensure] that none of the [muslims] should see them or associate with them, nor were they to speak with them even a word, and that they [i.e., the Mongol contingent] were not to travel except at night."[116] Despite the heavy security, Shīrāzī tells us of his success, in Cairo, of locating several much needed books for his commentary on Avicenna's *Canon* (as we will see in Chap. 3). A loosening of security once the embassy was in Cairo seems highly unlikely, and it is therefore not clear exactly how Shīrāzī was able to obtain his beloved books.

Of the mission's return 'Abd al-Ẓāhir states that the embassy headed first to Aleppo, "reaching it on the sixth of Shawwāl 681 (Jan. 7th, 1283 C.E.), and from there, headed back to its own country."[117] News of Tegüder Aḥmad's death (caused by dynastic struggles, on the 26th of Jumāda I, 683 A.H./Aug. 10th 1284)[118] arrived at Cairo during a second Ilkhan embassy. That embassy did not include Shīrāzī, but it was headed by Sheikh 'Abd al-Raḥmān himself.[119] (In addition the second embassy included four dervishes "for the sake of chanting," at which 'Abd al-Ẓāhir expresses his astonishment and wonder.)[120] According to 'Abd al-Ẓāhir it was the

[113]P.M. Holt, "The Īlkhān Aḥmad's Embassies to Qalāwūn: Two Contemporary Accounts," *Bulletin of the School of Oriental and African Studies, University of London* 49, no. 1 (1986): 128–132.

[114]Adel Allouche, "Teguder's Ultimatum to Qalawun," *International Journal of Middle East Studies* 22, no. 4 (November 1990): 437–446.

[115]Muḥyi al-Dīn Ibn 'Abd al-Ẓāhir, *Tashrīf al-ayyām wa al-'uṣūr fī sīrat al-malik al-manṣūr*, 1st ed. (al-Qāhirah: Wizārat al-thaqāfah wa-al-irshād al-qawmī, al-idārah al-'āmmah lil-thaqāfah, 1961), pt. 2. 5.

[116]Muḥyi al-Dīn Ibn 'Abd al-Ẓāhir, *Tashrīf al-ayyām wa al-'uṣūr fī sīrat al-malik al-manṣūr*, pt. 2. 5.

[117]Muḥyi al-Dīn Ibn 'Abd al-Ẓāhir, *Tashrīf al-ayyām wa al-'uṣūr fī sīrat al-malik al-manṣūr*, pt. 2., 16.

[118]Jackson, "AḤMAD TAKŪDĀR," Encyclopaedia Iranica, Encyclopaedia Iranica Online, 1984, http://www.iranica.com/articles/ahmad-takudar-third-il-khan-of-iran-r.

[119]Ibn 'Asākir, *Šāfi' Ibn 'Alī's Biography of the Mamluk Sultan Qalāwūn*, 328.

[120]Ibn 'Asākir, *Šāfi' Ibn 'Alī's Biography of the Mamluk Sultan Qalāwūn*, 329.

Mamluk sultan himself who conveyed news of Aḥmad's death to his sheikh, upon which the sheikh "fell into his arms, unconscious," dying shortly thereafter.[121]

In Rashīd al-Dīn's account of Aḥmad's rule, his rivalry with his nephew (and Abaqa's son), Arghūn, through which he ultimately lost his kingdom and his life, is an ever-present theme.[122] In Bar Hebraeus's *Chronography* we see Arghūn providing the following justification for the elimination of his uncle:

> Inasmuch as Aḥmad turned aside from the laws of our fathers, and trod the path of Islam, which our fathers did not know, all the princes agreed and they cast him forth from the kingdom, and sent him to the Khān, our great father, that he might judge him; and they seated me on the throne of the kingdom from the river Gihon to Frankistan.[123]

Given the skepticism with which many considered Tegüder Aḥmad's conversion to Islam, there is a fair amount of irony in this rationalization for Aḥmad's death.

2.3.4 A Return to Mongol Traditions: Arghūn (1284–1291 C.E.)

Like Abaqa, Arghūn had to await an official endorsement from Karakorum at his assumption to power,[124] and like him he had to contend with both the Golden Horde and the Chaghatai Khanate, his rivals to the north, and the east.[125] Though the purported proclamation by Arghūn in which he condemns Aḥmad Tegüder's conversion to Islam does not appear in Rashīd al-Dīn's history, his rule may have been characterized by a certain anti-Islamic sentiment (though some of what is reflected in the Muslim chronicles may be due to the Mongol tolerance of the various religions of their subjects). Upon assuming the throne Arghūn opted for non-Muslim viziers, first appointing Buqa, a Mongol notable, and subsequently Sa'd al-Daula who was Jewish.[126] Arghūn also appears to have forbidden the employment of Muslim scribes in the court bureaucracy.[127]

Arghūn's reign is also one in which Shams al-Dīn, the *Ṣaḥib Dīwān* under Hülegü, Abaqa, and Aḥmad, was put on trial and executed (Oct. 16th, 1284 C.E./Fourth of Sha'bān, 683 A.H.).[128] Already during the reign of Aḥmad, Arghūn had charged Shams al-Dīn and his brother with the poisoning of Abaqa. The

---

[121]Ibn 'Asākir, *Šāfi' Ibn 'Alī's Biography of the Mamluk Sultan Qalāwūn*, 332.

[122]Rashīd al-Dīn Ṭabīb, *Jāmi' al-tawārīkh*, 784–788.

[123]Bar Hebraeus, *The Chronography,* 474. Gihon is the Oxus River, from the Persian Jaiḥūn.

[124]Rashīd al-Dīn Ṭabīb, *Jāmi' al-tawārīkh*, 812.

[125]Rashīd al-Dīn Ṭabīb, *Jāmi' al-tawārīkh*, 821–822.

[126]Rashīd al-Dīn Ṭabīb, *Jāmi' al-tawārīkh*, 808.

[127]Jackson, "ARĞŪN KHAN," in *Encyclopaedia Iranica*, 1986, http://www.iranica.com/articles/argun-khan-fourth-il-khan-of-iran-r683-90-1284-91.

[128]Rashīd al-Dīn Ṭabīb, *Jāmi' al-tawārīkh*, 808–811; Bar Hebraeus, *The Chronography*, 472–473.

charge for which the great statesman was finally executed, however, was financial misappropriation.[129] ‘Alā’ al-Dīn Juwaynī, Shams al-Dīn’s brother and author of the *History of the World Conqueror*, had already died in 1283 C.E., likely from a stroke induced by the charges brought against him as a party, allegedly, to Abaqa’s death.[130]

Though a protege of Shams al-Dīn, Shīrāzī appears to have weathered the politics and intrigue of the court in this period and was even able to intercede for an acquaintance. We read about this in the first of two episodes recorded by Rashīd al-Dīn in which Shīrāzī appears in Arghūn’s presence. This episode belongs to sometime after the 13th of Jumāda al-ulā 689 A.H. (i.e., May 24, 1290):

> And at a post on the road to Vān, as the Sultan was returning from Alātāgh, Shīrāzī was received [in humility], and he made a presentation on the western sea and its harbors and its shores, which include many western and northern regions, and the king found his company to be exceedingly pleasant, as while recounting the regions of Rūm (Anatolia) the king had noticed Ammorium, which is in Rūm, and had asked Shīrāzī to describe it. He [i.e., Shīrāzī] presented a report of utmost eloquence containing prayers and plaudits for the king, and a description of the subject, which greatly impressed Arghūn. And as he was leaving for the hunt, he said to the Maulānā [i.e., Shīrāzī]: “When I return, come so that we may speak some more, for you speak wonderfully well.” He then pointed to Sa‘d al-Daula [the vizier] and indicated that they bring all three, meaning Amīrshāh, Fakhr al-Dīn Mustaufī, and the son of Hajjī Laylī, for they had taken all three from Rūm and had brought them. And Maulānā Shīrāzī reproached Sa‘d al-Daula in regard to Amīrshāh, and hastened him after the King, thus winning [Amīrshāh’s] release.[131]

We will meet Amīrshāh again in Chap. 3. The administrator of the loan taken by the Seljuk rulers from the Mongol treasury, Amīrshāh was also the dedicatee of the *Tuḥfa*, and thus a former patron of Shīrāzī. That Shīrāzī appears to have been able to chasten the vizier with respect to a prisoner and that he was even able to win the prisoner’s release indicates the extent of his authority during this period.

The second episode does not appear in the copy of the *Jāmi‘al-Tawārīkh* that was the main reference for this study.[132] It is included by Thackston in his translation of the *Jāmi‘al-tawārīkh* with a footnote stating that the text is absent from all manuscripts save a few.[133] The fragment which references Shīrāzī is quoted here from Thackston’s translation:

> [In addition to building, Arghūn] was also enthralled by alchemy, and alchemists came to his court from far and wide to encourage him in this art. Untold amounts of money were spent on it, but he never chided them for it and even cheerfully authorized more expenditures. One day an extremely subtle point was discussed in the presence of Maulānā Qutbuddin

[129]Biran, “JOVAYNI, ṢĀḤEB DIVĀN” in *Encyclopaedia Iranica*, 2009, http://www.iranica.com/articles/jovayni-saheb-divan.

[130]Lane, “JOVAYNI, ‘ALĀ’-AL-Dīn.” in Encyclopaedia Iranica, 2009, http://www.iranica.com/articles/jovayni-ala-al-Dn.

[131]Rashīd al-Dīn Ṭabīb, *Jāmi‘ al-tawārīkh*, 822–823. Shīrāzī would have been 55 years old.

[132]Rashīd al-Dīn Ṭabīb, *Jāmi‘ al-tawārīkh*, edited by *Bahman Karīmī* (Tehrān: Iqbāl, 1338)

[133]Rashīd al-Dīn Ṭabīb, *Rashiduddin Fazlullah’s Jami‘u’t-Tawarikh*, 577.

> Shīrāzī. When the alchemists had left, Arghūn said to the Maulānā [i.e., Shīrāzī], "Since I am only a Turk and you are a wise man, do you think these people are taking me for a ride? I have often wanted to put them to death, but since it is certain that this science exists and there must be someone who knows about it, if I withdraw my patronage from these ignorant men and put them to the sword, that one learned person will not trust me." In short, during Arghūn Khān's reign the alchemists spent untold amounts on their various experiments, but after much experimentation and tests, the veil of doubt was lifted from everyone's eyes, and nothing had been achieved other than financial loss and ruin.[134]

This episode to which this fragment refers is undated, appearing instead under the title "Part Three, on [Arghūn's] conduct and character; the pronouncements and orders he gave; incidents that occurred during his reign that were not included in the previous two sections but learned from various persons." The "lifting of the veil of doubt" in regard to alchemy could not have referred to the ruler himself, however, for in Rashīd al-Dīn's final chapter on Arghūn's life we see him still consorting with his alchemists:

> Arghūn Khān's belief in holy men and their customs was extremely strong, and he always sponsored and promoted that group. From India there came a holy man and claimed [the knowledge to] a long life. They asked him through what means is the life of holy men prolonged there? He said through a special draught. Arghūn asked him whether the draught was found locally. He said it was. [Arghūn] obliged the fashioning of it. The holy man produced a brew which contained Sulphur and Mercury. And he [i.e., Arghūn] partook of it for eight months at the end of which he spent forty days in seclusion in the fort of Tabrīz, and at that time no mortal was with him, except Orduquya and Qūcān, and Sa'd al-Daula, and the holy men who were constantly present and busy discussing their beliefs. When he left seclusion he decamped for Alātāgh and there an ailment appeared suddenly upon his humors, and Khwāja Amīn al-Daula, who was the physician at court, exerted himself, together with the other physicians, so that after a bit through their wise words some signs of health reappeared. [But] suddenly one day a holy man came and gave Arghūn three glasses of wine. Since he was still convalescing the illness returned and became terminal. And the doctors were unable to cure it and after two months of his sickness the generals started discussing and searching for the causes of his illness. Some said that the cause was the evil eye and that alms-giving was thus necessary, and some admitted that the shamans (who observed portents through the "art of the scapulae") were saying that the cause for the illness was sorcery and they placed the accusation on Tughanjūq Khātūn and through the beatings and the tortures of her trial they interrogated her and finally they drowned her and some other women. And this occurred on the 16th of Muḥarram of the year 690 A.H. [i.e., Jan. 19th, 1291 C.E.], and the Lord knows the truth of things.[135]

According to Rashīd al-Dīn, Arghūn commenced on taking the draught c. Ramadan of 688 A.H. (September 1289 C.E.): "On the fourth of Ramadan of 688 Arghūn Khān arrived at Marāgha and toured the observatory – and he commenced on drinking the black drug that will be described henceforth at that location [i.e., at Marāgha]. He then left for the cold-weather camp at Arrān."[136] It is difficult to know what to make of this tantalizing fragment, other than to emphasize the clear

---

[134]Rashīd al-Dīn Ṭabīb, *Rashiduddin Fazlullah's Jami'u't-Tawarikh*, 577.

[135]Rashīd al-Dīn Ṭabīb, *Jāmi' al-tawārīkh*, 824.

[136]Rashīd al-Dīn Ṭabīb, *Jāmi' al-tawārīkh*, 821.

association of Marāgha with the alchemical draught. The passage quoted earlier with respect to Arghūn's patronage of the alchemists has an interesting parallel in the final chapter of Hülegü's life which raises, at least, the possibility that Hülegü may have dabbled with alchemically produced potions and their purportedly life-prolonging qualities, as well. Interestingly, Hülegü's account also includes as the setting of its preamble the observatory at Marāgha; a connection that was already noted by Sayılı:[137]

> When fall arrived, aiming for the warm-weather camp at Zarrīneh-rūd, [the river] which the Mongols call Jaghātū, he [i.e., Hülegü] arrived at Marāgha and exerted himself in the completion of the observatory. And he loved knowledge exceedingly, and would encourage scientists in the pursuit of the ancient sciences (*awā'il*) and he had assigned salaries to all, and had embellished his court with the presence of the scientists and learned men, and he was interested in the science of alchemy, and [thus was] keenly interested in this group [i.e., the alchemists]. They lit many flames and burnt many drugs and blew through many useless bellows, large and small, and they had constructed pots from the "clay of wisdom," yet the concoctions only benefited them as far as their breakfast and evening victuals. They were ineffective as far as transmutation was concerned but in dishonesty and duplicity they had miraculous powers. They were unable to fuse a single dinar, nor were they able to mould a single dirham, yet they scattered the stores of the workshop of Divine Power to a place of oblivion and nonexistence. So much was spent on their provisions, desiderata, and stores that Qārūn himself ... had not been able to produce during his entire life [i.e., through the use of *his* elixir][138]

We will discuss the possibility of the presence of a Taoist tradition of alchemy at Marāgha in Chap. 6. Here we note that, if Hülegü's death, which as we saw involved the sudden return of symptoms such as weakness and an undefined ailment "upon his body" (a rash, perhaps?), was due to the ingestion of mercury or other toxic substance, then the irony of Rashīd al-Dīn's observations in regard to the wastefulness of alchemy is further amplified. As it is, Rashīd al-Dīn's account indicates that Arghūn almost certainly succumbed to voluntary poisoning, and that Hülegü may very well have done the same.

2.3.5 Culminating Crisis: Gaykhātū (1291–1295 C.E.) and Bāydū (1295 C.E.)

Subsequent to Arghūn death, it was his brother Gaykhātū who succeeded him. As with his uncle, Tegüder Ahmad, the beginning of his reign triggered a crisis of succession. The rival claimant in this case was Bāydū, Gaykhātū's cousin; and

[137]Sayılı, *The Observatory in Islam and Its Place in the General History of the Observatory*, 193.

[138]Rashīd al-Dīn Ṭabīb, *Jāmi' al-tawārīkh*, 734. For Qārūn see MacDonald, D.B. "Ḳārūn." *Encyclopaedia of Islam*, Second Edition. Brill Online, 2013. http://referenceworks.brillonline.com/entries/encyclopaedia-of-islam-2/karun-SIM_3951.

Hülegü's grandson through his fifth son, Taraqai.[139] Gaykhātū, as he appears in Bar Hebraeus and other historians, was a dissipated monarch given to debauchery with minors, forcing many of the Mongol nobility to send their children away to outlying districts.[140] His short reign included a military campaign to Anatolia, but none against the Mamluks.[141]

Rashīd al-Dīn refers to Gaykhātū's introduction of paper money, at the instigation of his *Ṣaḥib Dīwān* Ṣadr al-Dīn Zanjānī and other courtiers, as an "account of the inauspicious *chau*." Describing Gaykhātū's endorsement of this plan, Rashīd al-Dīn writes:

> [Since] Gaykhātū was an extremely liberal (sakhī') monarch and gave liberally [so that] the wealth of the entire world could not satisfy his generosity, he approved it.... And on the Monday of the nineteenth of Shawwāl of 693 A.H., they presented and set into circulation the *chau* in Tabriz, and it had been decreed that whoever would not accept it would be executed instantly. For a week they took it, fearful of the sword .... And most of the population of Tabriz had been forced to leave and goods and foodstuffs had been removed from the bazaar, so that nothing was left, and the people took refuge in the orchards, and a city of such dense population was utterly emptied of its people and the thugs and hooligans would strip of his belongings whomever they found in the streets.[142]

Rashīd al-Dīn writes that angered people mobbed a Quṭb al-Dīn "on a Friday in the congregational mosque."[143] Though not identified further, this Quṭb al-Dīn figure is almost certainly not our Quṭb al-Dīn but is rather the brother of Ṣadr al-Dīn Zanjāni (i.e., the mastermind behind the fiasco), who is identified as a chief judge in his own right, in the preceding chapter of the chronicle.[144] The experiment with paper money was a miserable failure, and appears to have petered out on its own once officials determined that it was unworkable.[145] Gaykhātū's rule did not outlive this fiasco by

---

[139]D. Morgan, *The Mongols*, 2nd ed. (Malden, MA: Blackwell Publishing, 2007), 225; Rashīd al-Dīn Ṭabīb, *Jāmi' al-tawārīkh*, 681.

[140]Bar Hebraeus, *The Chronography*, 494; B. Spuler, "Gaykhātū," in *Encyclopaedia of Islam*, Second Edition, Edited by: P. Bearman. (Brill Online, 2010), http://www.brillonline.nl/subscriber/entry?entry=islam_SIM-2427. *Encyclopaedia of Islam*, Second Edition. Edited by: P. Bearman, (Brill Online, 2010) http://www.brillonline.nl/subscriber/entry?entry=islam_SIM-2427.

[141]Spuler, "Gaykhātū." in *Encyclopaedia of Islam*, Second Edition. Edited by: P. Bearman. Brill Online, 2010. http://www.brillonline.nl/subscriber/entry?entry=islam_SIM-2427.

[142]See Rashīd al-Dīn Ṭabīb, *Jāmi' al-tawārīkh*, 835; Also Bar Hebraeus for the "immeasurable liberality of hand" which appears to be connected to his dissipated lifestyle (i.e., a lack of moral discipline in conduction with a lack of fiscal discipline) "'Whosoever hath in his hand silver, and doth not carry it to the offices of the Government to be stamped therein with [the word] Shaw, and giveth it up and taketh [in exchange] Shaw shall die the death.' And thus men remained in a state of great tribulation and indescribable difficulty for a space of two months." Bar Hebraeus, *The Chronography*, 496.

[143]Rashīd al-Dīn Ṭabīb, *Jāmi' al-tawārīkh*, 836.

[144]Ibid. 833; Rashīd al-Dīn Ṭabīb; At least one modern translation identifies this Quṭb al-Dīn with our Quṭb al- Dīn Shīrāzī, *Rashiduddin Fazlullah's Jāmi' u't-Tawarikh*, 808.

[145]Rashīd al-Dīn Ṭabīb, *Jāmi' al-tawārīkh*, 836; Bar Hebraeus, *The Chronography*, 496.

long. He was forced to deal with an insurrection by Bāydū that ultimately ended his rule. He was executed on Thursday, the Sixth of Jumāda al-ūlā of 694 A.H. (March 24th, 1295 C.E.).

The reign of Gaykhātū is not particularly relevant to our study of Shīrāzī. Indeed, as we will see in Chap. 3, the reign of Gaykhātū (together with the very brief reign of Gaykhātū's successor, Bāydū) is the only era during Shīrāzī's adult career in which there does not exist any evidence for the presence of Shīrāzī at the Ilkhan court.

Probably as a mark of loyalty to his employer, Rashīd al-Dīn includes the account of the short reign of Ghāzān's rival, Bāydū, in the chapter devoted to Ghāzān himself.[146] Since the account is of recent historical events the narrative achieves a level of detail that is lacking in earlier chapters. Rashīd al-Dīn's narrative of Bāydū culminates with his capture by the capable general Naurūz, roughly six months after taking the reigns of power. Upon hearing his request for a private audience, Ghāzān (Hülegü's grandson through Arghūn) requests instead that Bāydū be "finished off where he is,"[147] with the execution occurring in the "evening on Wednesday, the twenty third of Dhū al-Qa'da, 694 A.H. [Oct. 4, 1295 C.E.]."[148]

2.3.6 Reformation and Recovery: Ghāzān (1295–1304 C.E.)

Ghāzān is generally recognized for reversing the ruinous policies of his predecessor Ilkhanid rulers. Rashīd al-Dīn's *Jāmi'al-tawārīkh*, which includes within it some of Ghāzān's reform-minded proclamations, is the authoritative historical source for his reign. Ghāzān's reforms included a restructuring of the taxation system, a repeal of the expectation that Ilkhanid subjects provide quarters for traveling military and official personnel, a limiting of the burden on the Ilkhanid subjects of providing carriage animals for the governmental business, as well as other measures.[149] Morgan and others have pointed out that Rashīd al-Dīn was not an impartial observer in regard to his employer,[150] and it is certainly not surprising that Rashīd al-Dīn would have exaggerated the beneficence of Ghāzān, as well, perhaps, as the abuses perpetrated by his forebears. However, the reforms by Ghāzān of the exploitative system of taxation (which – as the *Jāmi'al-tawārīkh* fragment at the beginning of the

---

[146]Rashīd al-Dīn Ṭabīb, *Jāmi' al-tawārīkh*, 883.

[147]Rashīd al-Dīn Ṭabīb, *Jāmi' al-tawārīkh*, 915.

[148]Rashīd al-Dīn Ṭabīb, *Jāmi' al-tawārīkh*, ibid. See also, Barthold, W., "Baydu," in *Encyclopaedia Iranica*, 1988, http://www.iranica.com/articles/baydu-baidu-on-coins-badu-a-son-of-taragay-and-grandson-of-hleg-hulagu-reigned-as-il-khan-in-iran-from-joma.

[149]Petrushevsky, "The Socio-Economic Condition of Iran Under the Il-Khans," 495.

[150]Morgan, D., "Rashīd al- Dīn Ṭabīb." *Encyclopaedia of Islam*, Second Edition. Edited by: P. Bearman, (Brill Online, 2010) http://www.brillonline.nl/subscriber/entry?entry=islam_SIM-6237.

chapter indicates – had driven entire regions into ruin) were effective in salvaging the plight of the Ilkhanid subjects (and of the peasants, especially) – as can be seen in the appreciable rise of agricultural production during his reign.[151]

As we have noted Ghāzān's conversion to Islam is a topic to which Rashīd al-Dīn's devotes a considerable amount of space. An unfortunate side-effect with Ghāzān's conversion to Islam, however, appears to have been the reversal of the decades-long Ilkhanid policy of tolerance for the various religious practices of their subjects: "And on Wednesday the twenty-fourth of Dhū al-Qa'da, of the year 694 A.H. [Oct. 4, 1295 C.E.] it was proclaimed that in the capital Tabrīz, and in Baghdād and the other regions of Islam all of the temples of the shamans and the Buddhists and the churches and the synagogues be destroyed."[152]

Ghāzān's accession was complicated by rebellions that, at their root, were due to the crisis of succession at the end of Gaykhātū's reign. The situation appears to have taken several years to sort out, and was only settled after the execution of a rather long list of claimants to the throne. Also significant were a series of rebellions in Rūm (Anatolia), several of these by the Mongol overseers themselves (who were aided by various local factions). These were dealt with by Ghāzān by 1299 C.E.[153] The Seljuks of Rūm, in whose polity Shīrāzī had lived for some years, disappeared from the historical record in the first years of the following century, outliving these final spasms of violence by a handful of years, at most. Cahen notes the curious nature of the disappearance of the once powerful Seljuks of Rūm by stating that the "Sultanate disappeared in a manner so obscure that contemporaries do not mention it and authors who tried to account for it in retrospect disagree in regard to both dates and facts."[154]

---

[151] Petrushevsky, "The Socio-Economic Condition of Iran Under the Il-Khans", 495–496.

[152] Rashīd al-Dīn Ṭabīb, *Jāmi' al-tawārīkh*, 908; Waṣṣāf al-Ḥazrat, *Taḥrīr-i tārīkh-i Waṣṣāf*, 223. That traditional Mongol beliefs and practices outlasted this forceful top-down conversion effort can be seen, however, in an episode that appears in Kāshānī's history of Ghāzān's successor, Öljeitü. Of particular interest are several episodes in the year 709 A.H. (1309–1310 C.E.). A heated debate between the supporters of the Ḥanafī and Shāfi'ī schools in the court of Öljeitü appears to have been particularly vexing to the ruler. Öljeitü, who was born in 680/1282 and thus presumably followed Buddhism and the shamanism of his ancestors, not converting to Islam until the accession of his father, when he was 15 – appears to have cut short his audience by storming out. Subsequently, high-ranking officials had complained audibly for the good old peaceful days of the Mongol *yasa* system. 'Abd Allāh ibn 'Alī Kāshānī, *Tārīkh-i Ūljāyatū, Tārīkh-i pādishāh-i sa'īd Ghīyāth al-dunyá va al-Dīn Uljāyitū Sulṭān Muḥammad* (Tehrān: Bungah-i Tarjumah va Nashr-i Kitāb, 1348), 96. Tarjumah va Nashr-i Kitāb, 1348), 96. In the same year a lightning strike killed several courtiers, in the presence of the frightened ruler, forcing him to reconsider his religious convictions. "The amirs conveyed [to the Ilkhan] that according to the old conventions and the yasa of Chingiz Khan [he should be cleansed by fire]. They assembled the shamans who were in charge of this and said: 'This frightful lightning and incendiary and ruinous bolt is due to the ill omen of Islam and muslims. Should the King abandon the daily prayers and the adhan recital ... his passing through fire would be successful.'" Kāshānī, *Tārīkh-i Ūljāyatū*, 98.

[153] Cahen, *Pre-Ottoman Turkey*, 300–301.

[154] Cahen, *Pre-Ottoman Turkey*, 301.

Ghāzān's war against the Mamluks includes the military campaign of 1299 C.E./699 A.H. in which the Mongols were victorious, and temporarily occupied Damascus.[155] A final campaign against the Mamluks, in 1303 C.E./702 A.H., however, resulted in a decisive defeat of the Mongols.[156]

On the cultural front, it was Ghāzān who commissioned Rashīd al-Dīn to compose his history.[157] Waṣṣāf also mentions his construction of an observatory in Tabrīz, as part of a large complex that was started in 697 A.H. (1297/1298 C.E.) and finished in 702 A.H. (1302/303 C.E.).[158]

2.3.7 A Peaceful Interlude: Öljeitü (1304–1316 C.E.)

Though Rashīd al-Dīn was alive during the reign of Öljeitü and appears to have written a history of his reign, this history has not survived.[159] Our main sources for the reign of this ruler are instead Kāshānī's *Tārikh-i Oljaitu*, Waṣṣāf's history, as well as histories by Qazwīnī, and Banākatī.[160] It is through Kāshānī's text that we learn of Öljeitü's siege of the fort of Rahba on the Western bank of the Euphrates, in April of 1313 C.E.[161] This event, which was instigated by a group of renegade Syrian amirs, was to be the last Ilkhanid expedition against their arch-enemies, the Mamluks.[162] Despite this military campaign, which appears to have been a short and inconclusive affair and a 1314 C.E. conflict with the Chaghatai army in the east, Öljeitü's reign was generally speaking a peaceful one.[163]

---

[155] It is not known with certainty why the Mongols subsequently abandoned Syria, only to make a second unsuccessful attempt to retake it in the winter of 1300 C.E./700 A.H.

[156] R. Amitai, "ḠĀZĀN KHAN, MAḤMŪD," in *Encyclopaedia Iranica*, Iranica Online, 2000, http://www.iranica.com/articles/gazan-khan-mahmud.

[157] D. Morgan, "Rashīd al-Dīn Ṭabīb," in *Encyclopaedia of Islam*, Second Edition, Edited by: P. Bearman. (Brill Online, 2011), http://www.brillonline.nl/subscriber/entry?entry=islam_SIM-6237.

[158] Waṣṣāf al-Ḥazrat, *Taḥrīr-i tārīkh-i Waṣṣāf*, 229.

[159] Morgan, "Rashīd al-Dīn Ṭabīb."

[160] 'Abd Allāh ibn 'Alī Kāshānī, *Tārīkh-i Ūljāyatū*; Waṣṣāf al-Ḥazrat, *Geschichte Wassaf's*; Waṣṣāf al-Ḥazrat, *Taḥrīr-i tārīkh-i Waṣṣāf*; Ḥamd Allāh Mustaufī Qazwīnī, *Tārīkh-i Guzīdah*; Dāwūd ibn Muḥammad Banākatī, *Tārīkh-i Banākatī = Rawḍat ūlā al-albāb fī ma'rifat al-tawārīkh va al-ansāb* (Tehrān, 1348).

[161] Kāshānī, *Tārīkh-i Ūljāyatū*, 143.

[162] Kāshānī, *Tārīkh-i Ūljāyatū*, 143; D. Morgan, "Öldjeytü," in *Encyclopaedia of Islam*, Second Edition, Edited by: P. Bearman. (Brill Online, 2010), http://www.brillonline.nl/subscriber/entry?entry=islam_SIM-6018. *Encyclopaedia of Islam*, Second Edition. Edited by: P. Bearman, (Brill Online, 2010) http://www.brillonline.nl/subscriber/entry?entry=islam_SIM-6018.

[163] Morgan, "Öldjeytü." *Encyclopaedia of Islam*, Second Edition. Brill Online, 2010. http://www.brillonline.nl/subscriber/entry?entry=islam_SIM-6018.

Of particular relevance to our discussion is a fascinating episode in Öljeitü's career that involved a military campaign against the region of Gīlān.[164] This episode is remarkable due to the fact that Gīlān was located virtually at the heart of the Ilkhanid realms. That the region would require pacification a half-century after the arrival of Hülegü in Persia is, therefore, something of a paradox.[165] Though this episode appears in a number of Persian and Mamluk sources the details are not clear. It appears as though the campaign ended with a disastrous defeat of the Mongols, forcing the Persian sources (who were generally loyal to their Ilkhan overlords) to whitewash this uncomfortable fact.[166] The geography of the region – as characterized both by the rugged topography of the Alburz range, and by its heavy annual rainfall – was no doubt a factor in the defeat of the Mongols. This episode, dimly captured in the historic sources, is mentioned here because one of the local rulers of Gīlān, Amīra Dabāj, who appears briefly in these accounts, is the dedicatee of Shīrāzī's encyclopedic work the *Durra*. The significance of this fact for our study of Shīrāzī's life is discussed in Chap. 3.

It should also be noted here that Öljeitü was responsible for moving the capital city from Tabrīz, where it had been from the time of Abaqa, to the town of Sulṭānīya. Öljeitü's mausoleum, recognized as a supreme instance of Persian architecture, still stands in Sulṭānīya, where it was once part of a large religious complex.[167] It thus appears as though Shīrāzī was to live the last portion of his life a distance away from the politics and the hustle and bustle of the capital. If the accounts of Shīrāzī's sufism are to be believed, this likely would have been a welcome change for him.

2.3.8 *The Waning Years: Abū Sa'īd (1316–1335 C.E.)*

Coming to power after the death of his father in 1316, Abū Sa'īd was the last of the Ilkhanid line to rule Persia. His death in November 30 1335 C.E., which may have been by poisoning,[168] precipitated a crisis of succession and a prolonged power struggle.[169] That his death marked the end of an era can be seen from the fact that the historical records suddenly fall silent about the details of these power struggles

164 Charles Melville, "The Īlkhān Öljeitü's Conquest of Gīlān (1307): Rumour and Reality," in *The Mongol Empire and its Legacy*, Reuven Amitai-Preiss & David O. Morgan (eds.). (Leiden: Brill, n.d.), 73–125; Ḥamd Allāh Mustaufī Qazwīnī, Tārīkh-i *Guzīdah*, 607; Kāshānī, Tārīkh*h-i Ūljāyatū*, 55–71.

165 Kāshānī mentions the ruler of Gīlān as having payed homage to Hülegü upon the Mongol rulers arrival in Persia. Kāshānī, *Tārīkh-i Ūljāyatū*, 57.

166 Melville, "The Īlkhān Öljeitü's Conquest of Gīlān (1307): Rumour and Reality," 118.

167 Minorsky, "Sulṭānīya." *Encyclopaedia of Islam*, Second Edition. Brill Online, 2010. http://www.brillonline.nl/subscriber/entry?entry=islam_COM-1118.

168 Abu Abdallah Ibn Battutah, *The Travels of Ibn Battutah* (London: Picador, 2002), 78.

169 Morgan, *Medieval Persia, 1040–1797*, 79.

in which the protagonists were soon, in Boyle's words, so insignificant "that we are not even informed as to the time and manner of their death."[170] Thus the rule of the Ilkhanid dynasty ended with a whimper that was a faint echo of the demise of their vassals, the Seljuks of Rūm three and a half decades earlier.

As we saw Quṭb al-Dīn died 5 years prior to the accession of Abū Sa'īd and so the history of the Abū Sa'īd's reign is not directly relevant to our discussion. It should also be noted here, however, that it was during the reign of Abū Sa'īd that the great statesman and extraordinary historian Rashīd al-Dīn, who, along with Juwaynī, has left us the most important and detailed chronicles of the period, finally succumbed to the intrigue of the Ilkhanid court and was executed. His charge was the poisoning of Öljeitü.[171]

2.4 The Mongols and the Patronage of the Sciences

Having briefly reviewed the dynastic history of the Ilkhans and of their Mongol forbears in Persia I will now provide a provisional interpretation of the historical record in regard to the patronage of the sciences and especially of astronomy in this period. While recognizing the violence of the original campaigns early in the 13th century (a cataclysm that led not only to the demise of entire cultures in Central Asia but is linked, as well, to the extinction of certain cultural traditions such as the production of textiles in eastern Persia and the complete disappearance of *mīnā'ī* ceramics, for example)[172] many modern studies on the Mongols point out the culturally productive conditions of the subsequent decades: the patronage of luxury goods, the facilitation of trade across the Asian landmass along with the concomitant diffusion of new ideas of governance and religion, as well as the diffusion of various technologies related to arts and crafts through the relocation of artisans. Though the situation with science and scholarship is not clear, these enterprises presumably would have experienced a fate similar to that of other cultural traditions of the afflicted regions. It shouldn't be surprising, in other words, if certain scholarly and scientific traditions of the eastern Islamic world did not survive the conflagration (that had had, as we saw, the wholesale destruction of a fair number of urban centers as one of its defining characteristics), while others managed to survive and perhaps even to thrive in the culturally conducive factors listed above.

It perhaps bears pointing out here that the region afflicted by the military campaigns of the Mongols was one with a distinguished cultural tradition. When

[170] Boyle, "Dynastic and Political History of the Īl-khāns," 416.

[171] Abbas Iqbal, *Tārīkh-i Mughūl: az ḥamlah-'i Changīz tā tashkīl-i dawlat-i Taymūrī*, 6th ed. (Tihrān: Amīr Kabīr, 1365), 328.

[172] Linda Komaroff, "Introduction: On the Eve of the Mongol Conquest," in *The Legacy of Genghis Khan: Courtly Art and Culture in Western Asia, 1256–1353* (New York: Metropolitan Museum of Art, 2002), 4–5.

the last of the Chingiz's armies withdrew from Persia in 1226, the formerly bustling population centers that, according to the historical record, had been transformed to grizzly killing fields on an unimaginable scale (as we saw in the case of Balkh, Harāt, Marw, Nīshāpūr, Ṭūs) were many of the same that in earlier centuries had nurtured some of the luminaries of Islamic culture. A discussion of the factors that had led to the amazing military success of the Mongol armies is not within the scope of this study.[173] It is, however, worth remembering that had the conditions that allowed for the blinding success of Chingiz Khan and his army coalesced two centuries earlier, the resulting disruptions would have been contemporaneous with the lives of such luminaries as Bīrūnī, Ghazālī (Algazel), Ibn Sīnā (Avicenna), Rāzī (Rhazes), and Khayyām. While the centuries leading to the thirteenth century C.E. do not appear to have been particularly peaceful, one can wonder at the effect on the productive cultural milieu in which these well-known scholars were born and raised, had the Mongol war machine – with its habitual razing of urban centers – made an earlier appearance.[174]

There are, needless to say, factors that complicate a study of the impact of the Mongol campaigns on the cultural and scientific production of the era; among them the compounding affect of earlier trends of warfare and strife (see the introductory section of this chapter) and the fact that the events themselves no doubt represent a partial obliteration of historical data that may be particularly difficult to reconstruct and interpret after a span of 800 years. In a study based on biographical dictionaries covering the eighth to the thirteenth century C.E. Bulliet observes a precipitous decline in the scholarly activities of Persian scholars in the early decades of the eleventh century. This decline is therefore considerably earlier than the thirteenth century, and has ultimately been linked by Bulliet to environmental factors that affected the lucrative cotton crop of Persia.[175]

It is hoped that in due course enough studies are carried out on the surviving manuscripts themselves (both of the Mongol and preceding eras) to enable scholars to form a concrete picture of how various traditions of scholarship were transformed by the military campaigns of the Mongols under Chingiz Khan. In Chap. 5 the work of the great historian Ibn Khaldūn will be examined briefly, and his comments on

---

[173]See, for example, Morris Rossabi, "The Mongols and Their Legacy," in *The Legacy of Genghis Khan: Courtly Art and Culture in Western Asia, 1256–1353* (New York: Metropolitan Museum of Art, 2002), 15.

[174]In what can only be seen as a testament to the quality of scientific production in Persian-speaking lands in both the era leading to the Mongol conquests as well as the subsequent period, Kennedy, dubs the scientists of the Seljuk and Mongol periods as the "best of their age." See E.S. Kennedy, "The Exact Sciences in Iran Under the Saljuqs and Mongols," in *Cambridge History of Iran* (Cambridge: Cambridge University Press, 1968), 679. Sadly, the state of scholarship does not yet permit a conclusive determination of the impact of the Mongol invasions themselves.

[175]Richard Bulliet, "Abu Muslim and Charlemagne," in *Community, State, History and Changes: Festschrift for Prof. Ridwan al-Sayyid* (Beirut: Arab Network for Research and Publishing, 2011), 25–26. See also, Richard Bulliet, *Cotton, Climate, and Camels in Early Islamic Iran* (New York: Columbia University Press, 2009), 142.

Persian scientists will be used to suggest, at least, that the impact of the Mongol campaigns on the cultural production of the Persia was more apparent to medieval historians than they are to modern scholars.

A Mongol practice that has been cited as a factor for cultural productivity in periods subsequent to the original Mongol campaigns is that of the relocation of captives and slaves to faraway destinations. At several points in his narrative (written some 30 years after the original) Juwaynī describes the relocation of artisans (and occasionally of young women).[176] He states, as well, that some of the buildings at Karakorum were built with the assistance of "muslim" masons.[177] It is difficult, however – given the mayhem and chaos reflected in the historical narratives of the Mongol campaigns – to imagine a similarly perceived need to preserve the scholarly traditions of the conquered lands in western Asia. Some scholars would no doubt have been spared to act as interpreters and functionaries in the bureaucracy of the Mongol empire, especially in the Persian-speaking areas to the north and north-east of the Oxus river. Yet, it is safe to assume – given the historical data we have available to us – that these would have been the exceptions rather than the rule.

It is also not unreasonable to assume that during the era of the viceroys the scientists and scholars who survived the military campaigns of the Mongol armies, would have had greater concerns than the pursuit of their craft or the seeking of patronage for such pursuits. In Harawī's account of the aftermath of the fall of Harāt we read that a small number of survivors (20–40 souls) lived initially on "the flesh of humans and of dogs" and that for the subsequent 4 years they were forced to prey on passing caravans for survival.[178] Harawī also relates that "from the year 619 A.H. to 634 A.H. (i.e., 1222/1223 to 1236/1237 C.E.) the city was a ruin; so that in these fifteen years no creature lived here, other than the occasional brigands [singular, *ayyār*] who were either in Harāt or in the nearby foothills."[179] Under these conditions it is likely that the scientists who had survived the campaigns and who had the ability would have sought refuge and patronage in well-defended locations,

[176] Juwaynī, *Genghis Khan*, 107; Sayf ibn Muḥammad ibn Ya'qūb, *The Ta'rikh Náma-i-Harat (The History of Harát) of Sayf Ibn Muḥammad Ibn Ya'qúb Al-Harawí*, 81.

[177] Juwaynī, *Genghis Khan*, 237.

[178] Sayf ibn Muḥammad ibn Ya'qūb, *The Ta'ríkh Náma-i Harat (The History of Harát) of Sayf Ibn Muḥammad Ibn Ya'qúb Al-Harawí*, 81–90. Harawī describes the transformation of the once-bustling metropolis of a hundred-thousand souls to an eerie moonscape as follows: "And in these four years, the few places in the city that had remained undamaged collapsed by virtue of the falling of the rain and the density of the snow, and the city became a place of such [terror] it was as though at each rest a ghoul [was hiding] or at each step [one could hear] a keening wail." In the same source we read that, as Chingiz Khan had followed a scorched earth policy, "from the environs of Balkh to Damghan people ate the flesh of humans, dogs and cats for one year." This indicates that the campaigns managed to blight not merely the cities that had been targeted militarily but to destroy the entire countryside as well, as the agricultural systems of the whole region appear to have collapsed. Sayf ibn Muḥammad ibn Ya'qūb, *The Ta'ríkh Náma-i Harat,* 87.

[179] Sayf ibn Muḥammad ibn Ya'qūb, *The Ta'ríkh Náma-i Harat*, 93.

as in the case of Ṭūsī who, in this period, found refuge with the Ismailis and their virtually impregnable forts.[180]

To imagine the pace of the recovery during the reign of the viceroys (i.e., the three decades separating the withdrawal of Chingiz and the arrival of his grandson Hülegü) and of the Ilkhans, we need only note that by the time Qazwīnī was writing his *Nuzhat al-qulūb* during the reign of Ghāzān (i.e. a little under a century after the original conflagration), of those destroyed cities that had been reconstructed many were rebuilt in a reduced scale: large towns were transformed into smaller towns or villages (and small towns to villages, etc.). Among the towns that were rebuilt in such reduced circumstances Qazwīnī lists a considerable number; we note here Qum, Sīrāf, Mīāneh and Kermānshāh as examples.[181] However, Qazwīnī is careful to point out as well that many of the towns (such as Khurrābād, Saimara, Arrajān, and Dārābjird) were still in ruins in his time, nearly a century after their destruction.[182] Indeed, some of the major population centers of medieval Persia– Rayy,[183] Marw,[184] Balkh,[185] notable among them – were left as ruin-fields for many

[180]Mudarris Razavi, *Aḥwāl wa Athār-i Muḥammad Ibn Muḥammad Ibn al-Ḥasan al-Ṭūsī*, 4. Several historical sources state that Ṭūsī was held by the Isamili's against his will. Ibid. Certainly anti-Ismaili factionalism and the desire to rationalize Ṭūsī's long stay with the Ismailis should be accounted for when interpreting these accounts. In the conclusion to his commentary on Avicenna's Kitāb al-ishārāt wa al-tanbīhāt (or "Book of Directives and Remarks"), which was completed in the middle of Ṣafar, 644 A.H. (c. the beginning of July, 1247 C.E.) Ṭūsī speaks of "having written the majority of the book in such straitened circumstances, that it would be impossible to imagine worse." Razavi interprets this as indicating Ṭūsī's difficulties with the Ismailis. In my mind the reference could be to the desolation induced by the war, for he also writes: "And [as for] the continuance of my life – its [military] ruler are my sorrows, and its soldiery are my anxieties." Mudarris Razavi, *Aḥwāl wa Athār-i Muḥammad Ibn Muḥammad Ibn al-Ḥasan al-Ṭūsī*, 7.

[181]Petrushevsky, "The Socio-Economic Condition of Iran Under the Il-Khans," 497; Ḥamd Allāh Mustaufī Qazvīnī, *The Geographical Part of the Nuzhat-Al-Qulub Composed by Ḥamd-Allāh Mustawfī of Qazwīn in 740 (1340).*

[182]Petrushevsky, "The Socio-Economic Condition of Iran Under the Il-Khans," 497.

[183]V. Minorsky, "al-Rayy," in *Encyclopaedia of Islam*, Second Edition, Edited by: P. Bearman. (Brill Online, 2011), http://www.brillonline.nl/subscriber/entry?entry=islam_COM-0916; Ruy González de Clavijo, *Narrative of the Embassy of Ruy Gonzalez De Clavijo to the Court of Timour at Samarcand, A.D. 1403–6: Translated for the First Time with Notes, a Preface, and an Introductory Life of Timour Beg* (New Delhi: Asian Educational Services, 2001), 99.

[184]A. Yu. Yakubovskii, "Marwal- SHāhidjān," in *Encyclopaedia of Islam*, Second Edition, Edited by: P. Bearman. (Brill Online, 2010), http://www.brillonline.nl/subscriber/entry?entry=islam_SIM-4978; See also González de Clavijo, *Embassy of Ruy Gonzalez De Clavijo*, 117.

[185]The Chinese Taoist monk Ch'ang-Ch'un was able to visit the ruins of Balkh in 1223, ibid. 487, as did Marco Polo (probably during the reign of Arghūn). See Marco Polo, The Travels of Marco Polo: The Complete Yule-Cordier Edition: Including the Unabridged Third Edition (1903) of Henry Yule's Annotated Translation, as Revised by Henri Cordier, Together with Cordier's Later Volume of Notes and Addenda (1920) (New York: Dover Publications, 1993), 151. Writing of his visit to Balkh in the fourteenth century, Ibn Battuta relates: "It is completely dilapidated and uninhabited, but anyone seeing it would think it to be inhabited because of the solidity of its construction (for it was a vast and important city), and its mosques and colleges preserve their outward appearance even now, with the inscriptions on their buildings incised with lapis-blue

centuries or abandoned permanently. Given the evidence of the historical record the impression can not be avoided that parts, at least, of the Persian-speaking world were transformed to virtual moonscapes or perhaps reconfigured into vast grazing fields for the herds of pastoralist conquerors. It is perhaps not surprising, then, that the decision to formally consolidate the Mongol holdings in Persia only happened in the sixth decade of the century. While internal factors involving politics of the Mongol rulers no doubt played a role, it was also perhaps the case that by this point enough of a recovery had taken place to make a full-scale occupation worthwhile in the first place.

As we have noted before, areas that were fortunate to not experience the Mongol armies directly would have felt the disruptions to a considerably lesser degree. Shīrāzī's home-province of Fārs was one such area. We will look at Shīrāzī's life in Chap. 3. Here we merely point out that as far as we can discern from the biographical material regarding Shīrāzī, his youth and his education do not appear to have been affected by the turmoil caused by the Mongols. Yet, as an intellectual and courtier Shīrāzī would have been frequently reminded of the political realities of his own era that had directly resulted from the trauma earlier in the century. There is little doubt that during his travels (particularly to Khurāsān) he would have witnessed first hand, the midden-heaps to which Juwaynī refers, and which would have been a constant reminder of the violent events that had so recently affected the region.

Möngke Khān's request that Ṭūsī be sent to Karakorum is from the end of the era of the viceroys. And it may be one of the earliest records of an attempt to preserve scientists from the Islamic world for the benefit of the Mongol rulers. That this incident has been preserved speaks no doubt of the great fame of Ṭūsī, but perhaps was also a signal of a heightened awareness by the Mongol rulers of the dependence of urban civilization on scholars as a practical matter. That men of letters had been prized earlier as administrators is demonstrated by 'Alā' al-Dīn Juwaynī's career at the Mongol court, but the case with Ṭūsī suggests that perhaps the project to attract the best scholarly "talent" of the far-flung Mongol empire to its center was widened at some point during the reign of the viceroys to include scientists as well. On the Great Khan's recruitment effort Rashīd al-Dīn writes:

> From among the kings of the Mongols, Möngke Qā'ān had been distinguished by great intelligence, perspicacity, and judgement, to the level that he had solved some of the problems of Euclid. His exalted will ... had obliged the building of an observatory. He appointed Jamal al-Dīn Muḥammad ibn Ṭāhir ibn Muḥammad al-Zaydī Bukhārī to carry out the project, yet some of the operational details were unclear to him, while at the same time the reputation of the superior learning of Ṭūsī had been as globe-traversing as the wind. At the time of leave-taking Möngke had asked his brother, as soon as the forts of the unbelievers had been taken, to send Khwājah Naṣīr al-Dīn back to Karakorum. Yet at the time [of the fall of the Ismaili forts], since Möngke Qā'ān was preoccupied with the

paints. The accursed [Chengiz] devastated this city and pulled down about a third of its mosque because of a treasure which he was told lay under one of its columns. It is one of the finest and most spacious mosques in the world; the mosque of Ribat al-Fath in the Maghrib resembles it in the size of its columns, but the mosque of Balkh is more beautiful than it in all other respects." Ibn Battutah, The Travels of Ibn Battutah (London: Picador, 2002), 144.

> conquest of the lands of the Manzī [i.e., in China] and was thus away from his throne, Hulākū decreed that he would build the observatory [in Persia] for he had become aware of [Ṭūsī's excellent qualities].[186]

Thus, according to Rashīd al-Dīn, the building of the Marāgha observatory appears to have been due to the Hülegü seizing an unexpected opportunity during his campaign of 1256 C.E.

Hülegü's campaign has been compared for its violence to the campaigns of Chingiz during 1216–1225 C.E.[187] This does not appear to be a fair comparison. While the historical record offers glimpses of the resistance against Hülegü's campaign (resistance that would no doubt have resulted in violent punitive measures), the intensity of the earlier campaigns and the wide geographical extent of the destruction are not reflected in the historical accounts.[188]

On the other hand, it is unlikely that Hülegü was a particularly benevolent ruler (as has been recently suggested by some historians of the Mongol period).[189] Thanks to the work of Petrushevsky and others, who have examined the historical evidence of agricultural production and tax revenues for Persia under Mongol rule it is possible to trace the precipitous economic decline of Persia in the thirteenth century subsequent to the invasion of the Mongols.[190] The exploitation of peasant farmers through arbitrary and often draconian taxation, and the heavy environmental impact of the great numbers of newly-arrived nomad pastoralists were factors that contributed to the onerous economic conditions of Persia during this period.[191] The declining trend of agricultural production continued through the Ilkhanid period and was only reversed at the end of the century during the reign of Ghāzān.

In discussing the social policy of the Ilkhans Petrushevsky identifies two competing processes within the Mongol aristocrats and the Persian elites allied to them: a process that aimed at "the creation of a strong central authority in the person of the Il-Khan and the adoption by the Mongol state of the old Iranian traditions of a centralized feudal from of government," as well as a trend that was

---

[186]Rashīd al-Dīn Ṭabīb, *Jāmi' al-tawārīkh*, 718.

[187]Komaroff, "Introduction: On the Eve of the Mongol Conquest," 3. In contrast, the effort to portray Hülegü as an enlightened warrior/ruler is a trend that has gained in popularity recently. One of the most active proponents of this revisionist school is George Lane; see *Genghis Khan and Mongol Rule* (Westport, Conn: Greenwood Press, 2004), 60–62.

[188]See Juwaynī, *Genghis Khan*, 615.

[189]One of the most active proponents of this revisionist school is George Lane; see *Genghis Khan and Mongol Rule* (Westport, Conn: Greenwood Press, 2004), 60–62.

[190]I. Petrushevsky, *Kishāvarzī va munāsabāt-i arzī dar Īran-i ahd-i Mughūl, Qarnhā-yi 13 va 14 mīlādī* (Tehran: Mu'assasah-'i Muṭāla'āt va Taḥqīqāt-i Ijtimā'ī, 1344); Ann K.S. Lambton, *Landlord and Peasant in Persia: A Study of Land Tenure and Land Revenue Administration* (London: I.B. Tauris, 1991); Petrushevsky, "The Socio-Economic Condition of Iran Under the Il-Khans"; See also Ann K.S. Lambton, *Continuity and Change in Medieval Persia: Aspects of Administrative, Economic, and Social History, 11th–14th Century* (Albany, N.Y.: Bibliotheca Persica, 1988).

[191]Petrushevsky, "The Socio-Economic Condition of Iran Under the Il-Khans," 490.

"antagonistic to settled life, agriculture and to towns," and supported "unlimited, rapacious exploitation of settled peasants and town-dwellers."[192] Writing of the second trend Petrushevsky states: "These representatives of the military feudal-tribal steppe aristocracy regarded themselves as a military encampment in enemy country, and made no great distinction between unsubjugated and subjugated settled peoples. The conquerors wished to plunder both ... the former by seizure of the spoils of war, the latter by exacting burdensome taxes. The supporters of this policy did not care if they ended by ruining the peasantry and the townspeople; they were not interested in their preservation. The most self-seeking and avaricious members of the local Iranian bureaucracy supported the adherents of this ... trend, as did the tax-farmers, who closely linked their interest to that of the conquerors and joined with them in the plunder of the settled population subjected to taxation – the *ra'yat*."[193] It appears as though it was this second group that predominated during the rule of Hülegü and his successors up to and including the short-lived reign of Bāydū. The enfeeblement of the economy that resulted from many decades of "rapacious" rule was no doubt one of the factors that ultimately forced the economic reforms of Ghāzān. Petrushevsky chronicles the enervated state of an economy teetering on the edge of collapse due to decades of depredation and misrule, tracing as well the positive effects of the policy shift under Ghāzān, for which he credits Ghāzān's chief administrator (and the eminent historian without whom the historical knowledge of the era would be greatly impoverished) Rashīd al-Dīn, himself.[194]

Given the generally predatory qualities of the era of Ilkhan rule, it is therefore somewhat ironic that we are able to recognize Hülegü as the instigator of one of the most important acts of scientific patronage in the medieval Islamic era: the construction of the Marāgha observatory. Though observatories had not been unknown in Islamic world prior to Marāgha,[195] the observatory at Marāgha, the building of which commenced shortly after the fall of Baghdād, was notable for its physical scale, the scope of its program, and its longevity relative to those that had gone before it.[196] To obtain a better sense of how this act of scientific patronage came about, it is useful to examine the events leading to Hülegü's involvement with this project.

As we saw in Rashīd al-Dīn's comments on Möngke, prior to setting off for Persia Hülegü was aware of his brother's plan for building an observatory in China. It is not clear, however, when he decided to build an observatory of his own, thus anticipating Möngke's project. At the fall of Alamūt, Juwaynī tells us of his visit to the

[192] Petrushevsky, "The Socio-Economic Condition of Iran Under the Il-Khans," 491.

[193] Petrushevsky, "The Socio-Economic Condition of Iran Under the Il-Khans," 492.

[194] Petrushevsky, "The Socio-Economic Condition of Iran Under the Il-Khans," 494–500.

[195] Samsó, J. "Marṣad." *Encyclopaedia of Islam*, Second Edition. Brill Online, 2010. http://www.brillonline.nl/subscriber/entry?entry=islam_SIM-4972.

[196] Sayılı, *The Observatory in Islam and Its Place in the General History of the Observatory*, 189–223.

library and observatory.[197] Hülegü is not mentioned in this account at all implying that, at this stage, the Mongol warlord was not yet preoccupied with the construction of an observatory. Yet, this situation appears to have changed on the way to Baghdād, suggesting that an adviser (perhaps Ṭūsī, himself) may have convinced Hülegü of the importance of the founding of an observatory in Persia, itself.[198] Indeed, the recruitment of al-'Urḍī (who as the builder of the instruments would have been one of the earliest members of the Marāgha observatory team) suggests that by the time of his Syrian campaigns (less than a year after the fall of Baghdād) Hülegü was committed to acquiring the best talent for his observatory. Though the precise circumstance of al-'Urḍī's trip to Marāgha are not known, al-'Urḍī himself writes that he was unhappy at Marāgha, for being away from his homeland and for being tasked with things that were not "within his main line of work."[199] The tone of frustration suggests that al-'Urḍī was taken to Marāgha against his will.

Indeed, the circumstances of al-'Urḍī's trip to Marāgha may have been similar to Muḥyi al-Dīn al-Maghribī's, whose professional capacities as an astronomer ensured that his life alone, from among those of his companions at the court of Malik Nāṣir at Damascus, was spared. Al-Maghribī's first-person account appears in Bar Hebraeus's history and in it he describes how he saved himself in the nick of time by announcing his profession during the course of an ambush by Mongol soldiers.[200] Al-Maghribī was subsequently sent to Marāgha, indicating again that Möngke's project for building an observatory and for recruiting scientific talent had by this stage been fully adopted by Hülegü himself.

In Rashīd al-Dīn's account of the founding of Marāgha, he credits Hülegü (albeit in vague terms) as the person responsible for the founding of the Marāgha observatory.[201] Yet, other accounts exist that explicitly credit Ṭūsī as the mastermind behind the Marāgha observatory. These accounts, though of a fabulous nature, are more consistent with the fact that at the outset the observatory project was not Hülegü, but Möngke's. The following anecdote in which Ibn Shākir attributes the founding of Marāgha to Ṭūsī appears in *Fawāt al-wafāyāt*:

> They say that when [Ṭūsī] desired to [build the observatory] Hulāgū saw what he was longing for, and so said to him: Of what use is this science that is related to the stars? Can what has been ordained be avoided? [Ṭūsī said:] I will show you an example: "[Order O Khān] someone to climb to that location and to throw from its top a large copper vessel without anyone knowing of it." So he did so. And when this occurred a great noise was created so that all who were present were terrified, some to the point of passing out, but as for Ṭūsī and Hulāgū, not a thing happened to them by virtue of their knowledge of what had occurred. So he said to him: "The science of the stars has this benefit: he who is conversant

---

197 Juwaynī, *Genghis Khan*, 719.

198 George Saliba, "Horoscopes and Planetary Theory: Ilkhanid Patronage of Astronomers," in *Beyond the Legacy of Genghis Khan* (Leiden: Brill, 2006), 357–368.

199 al-'Urḍī, *Kitāb al-hay'a*, 29.

200 Bar Hebraeus, *The Chronography*, 438.

201 Rashīd al-Dīn Ṭabīb, *Jāmi' al-tawārīkh*, 718.

> in it is aware of what is happening, so the fear that is created for the oblivious and the unaware [does not affect] him." So [Hülegü] said: "There is no harm in this," and ordered him to commence [in the building of the observatory]."[202]

Though there is no way to ascertain Hülegü's feelings on astrology we could perhaps speculate that his appreciation for this art was likely similar to the views of his grandson with respect to alchemy: a recognition of his own ignorance coupled with certainty as to the validity and the critical importance of the esoteric craft.[203]

It is certainly true that the belief about celestial bodies and how their influence suffused the sublunar realm was practically universal in the medieval world. It would thus be a mistake to dismiss the many references to fate and the workings of the celestial bodies in historical works of the period (such as Juwaynī's, for example) merely as figures of speech. In the introduction of his history, Juwaynī follows a declaration of the importance of patronage to literature and to scholarship, with a lamentation on the capriciousness of Fate (one of many that appears in his work):

> But because of the fickleness of Fate, and the influence of the reeling heavens, and the revolution of the vile wheel, and the variance of the chameleon world, colleges of study have been obliterated and seminaries of learning have vanished away; and the order of students has been trampled upon by events and crushed underfoot by treacherous Fate and deceitful Destiny.[204]

While using here some of the rhetorical flourishes that were common to an educated man of his cultural background, there is again little reason to doubt Juwaynī's underlying belief that inexplicable terrestrial phenomena (no doubt such as the cataclysm of the Mongol invasions themselves) were caused by the "influence of the reeling heaves."[205] The strategic role of the stars and their influence on the events in the sublunar world are also glimpsed in Rashīd al-Dīn's account of the accession of Abaqa, the date of which, as we saw was chosen by Ṭūsī. Elsewhere in Rashīd al-Dīn's history we find Hülegü in consultation with his newly acquired adviser in regard to the providential risks associated with his siege of Baghdād. Though Ṭūsī's astronomical knowledge is not explicitly part of the counter-argument he presents to those who opposed the campaign, it is not difficult to imagine how Ṭūsī's knowledge of the stars would have been an important part of his authority. Indeed, earlier in the same episode Hülegü asks another of his courtiers by the name of Ḥusām al-Dīn-i Munajjim (i.e., Ḥusām al-Dīn, the astrologer/astronomer) "who had escorted him by order of the Qā'ān [i.e., Möngke] – so as to choose the moment of his mounting and dismounting from his horse – to tell, without embellishment all the portents of the stars."[206] It is reasonably clear, therefore, that Hülegü's patronage

---

[202] Kutubī, *Fawāt al-wafāyāt wa al-dhayl 'alayhā*, 3, 247.

[203] See Saliba, "Horoscopes and Planetary Theory: Ilkhānīd Patronage of Astronomers" for a different interpretation of Hülegü's views on Marāgha.

[204] Juwaynī, *Genghis Khan*, 5. The translation used here is that of Professor Morgan.

[205] Waṣṣāf, *Geschichte Wassaf's*, 100.

[206] Rashīd al-Dīn Ṭabīb, *Jāmi' al-tawārīkh*, 706.

of the Marāgha observatory was due to its importance in the security, prosperity, and success of the ruler and (by extension) of the Ilkhanid state.[207] In examining the historical record one can not help wondering if the attention lavished upon the Marāgha observatory was not analogous to the care bestowed upon modern research centers that are engaged in the production and practice of cutting edge technology for the purpose of preserving the security and welfare of the state. If this view is accepted, then it is also reasonable to assume that a good fraction of the attention paid to astronomers and their research was due to the power of astronomy as a strategic tool for providing yearned-for and much needed knowledge regarding the impact of the reeling heavens on events and their circumstances on Earth.

[207]The situation is clearly similar with the patronage of the other scientific activity that garners multiple references in the historical sources: that of alchemy. This enterprise would have been viewed in connection to the granting of eternal life to the Ilkhān, as we saw in the episode of Abaqa's death, it would have been a particularly important recipient of royal patronage.

Chapter 3
Shīrāzī's Life

3.1 Introduction

A survey of the biographical information that has reached us in regard to Shīrāzī appears in two publications in Persian: Minovi's article in the festschrift honoring V. Minorsky, *Yādnāmeh-i Irāni-i Minorsky*, and a biography by Mir.[1] Much of this information has in turn been translated and expanded upon by Walbridge for use in his book *The Science of Mystic Lights*.[2] A brief glance at all three works indicates that in addition to certain coherent features of the various accounts a considerable amount of material has been added to Shīrāzī's lore by way of accretion in the years that have passed since his death. In this chapter I will review this surviving biographical information on Shīrāzī with an emphasis on the episodes that are presented with some consistency in the earliest surviving sources. For a comprehensive list of the reported events of Shīrāzī's life the reader is referred to the excellent works by Minovi, Mir, and Walbridge.

The sources for this chapter have been listed in Chap. 1 (Sect. 1.3.2). They include Shīrāzī's autobiography, and the information appearing in the works of Ibn al-Fuwaṭī (1244–1323 C.E./642–723 A.H.), al-Dhahabī (1274–1348 C.E./673–748 A.H.), al-Sallāmī (d. 1372 C.E./774 A.H.), and Ibn Ḥajar al-'Asqalānī (1372–1449 C.E./773–852 A.H.).

In the subsequent sections of this chapter Shīrāzī's autobiography will be examined first in an effort to identify the key episodes of his life. Material from the other sources listed will be added to both provide additional detail to Shīrāzī's

[1]Minovi, "Mulla Qutb Shirazi"; *Mir, Sharḥ-i ḥal wa āsār-i 'allamah Qutb al-Dīn Maḥmūd Ibn Mas'ūd Shīrāzi, danishmand-i 'ali qadr-i qarn-i haftum, (634–710 A.H.).*

[2]Walbridge, *The Science of Mystic Lights*; See also Walbridge, "The Philosophy of Qutb al-Din Shirazi."

K. Niazi, *Quṭb al-Dīn Shīrāzī and the Configuration of the Heavens: A Comparison of Texts and Models*, Archimedes 35, DOI 10.1007/978-94-007-6999-1_3,

account as well as to describe those episodes about which Shīrāzī is silent in his autobiographical notes. A comparison with the works by Minovi, Mir, and Walbridge suggests that these authors cover a great deal of what can reasonably be said about Shīrāzī's life.

3.2 Shīrāzī's Biographical Information in the *al-Tuḥfa al-Sa'dīya*

Shīrāzī's *al-Tuḥfa al-Sa'dīya* is a commentary on the first book of Avicenna's *Canon of Medicine*.[3] It is the only known work of Shīrāzī that has a biographical introduction, and it is thus likely that Shīrāzī considered it his major work. Minovi suggests that Shīrāzī wrote a commentary on the entirety of Avicenna's *Canon of Medicine*.[4] Walbridge doubts this is the case, suggesting that Shīrāzī's commentary is limited to the first book of the *Canon*, i.e., the *kullīyāt* or principles.[5] As we have said the fragment of Shīrāzī's introduction to his commentary has been reproduced nearly in its entirety in Mishkat's edition of the *Durra*. This is the edition that was generally used for the present study. The manuscript MS Suleimaniya 3649 was used to fill in the gaps for this text.[6]

Shīrāzī begins by giving a brief account of his family members and their experience in medicine:

> I was from a household that was famed in this art ... by virtue of my family's success in the treatment and the correction of the complexions with Jesus-like breathes and Moses-like hands, I too rejoiced, in the bloom of my youth, in attaining and comprehending it both in detail and in summary. And I engaged in all that was associated with medicine and with ophthalmology as far as the manual techniques such as venesections, extractions, "tucking" [i.e., blepharoplasty], conjunctival peritomies, pterygiectomies and others.... And all of this I did beside my father, Imam Ḍiya' al-Dīn Mas'ūd Ibn al-Muṣliḥ al-Kāzerūnī ... who was considered to be the Hippocrates of his age and the Galen of his day.[7]

[3]Walbridge, *The Science of Mystic Lights*, 186.

[4]Minovi, "Mulla Qutb Shīrāzī," 173.

[5]Walbridge, *The Science of Mystic Lights*, 186. It should be noted that in his introduction to *al-Tuḥfa al-Sa'dīya*, Shīrāzi only mentions the *Kullīyāt*, i.e., the first book of the *Canon*. So, while definitive proof is currently lacking, Professor Walbridge will likely be proven right, once a proper study of the *al-Tuḥfa al-Sa'dīya* is carried out. That a simple fact such as this remains unresolved is a telling comment on state of scholarship on Shīrāzī.

[6]Quṭb al-Dīn Shīrāzī, *al-Tuḥfa al-sa'dīya fī al-ṭibb,* Suleimaniya MS 3649. In addition a partial Persian translation of this text appears in Nurani's edition of the Sharḥ ḥikmat al-ishrāq. Quṭb al-Dīn Shīrāzī, *Sharh-i Hikmat al-Ishraq-i Suhravardi*, v–x.

[7]Shīrāzī, *Durrat al-tāj li-ghurrat al-dabāj*, kh. For the medical terms in this passage see *Albucasis on Surgery and Instruments*, by Abū al-Qāsim Khalaf ibn 'Abbās al-Zahrawī, M.S. Spink, and G.L. Lewis, Ed., (University of California Press, Berkeley, 1973), p. 212.

At his father's death, Shīrāzī who was still an adolescent was promoted to take his place:

> And since I had developed a reputation as one with good instinct and acumen I was made physician and ophthalmologist in the Muẓaffarī hospital in Shīrāz after the death of my father, when I was fourteen years old. And I stayed there for ten years as one of the doctors who did not desist from studies except to provide treatment ... for the reason that my soul was not satisfied with that which my contemporaries were content ... rather it drove me to exert my utmost in it so that I would attain the highest level of achievement.[8]

It was at this stage of Shīrāzī's career that he initiated a project that was to preoccupy him for the rest of his life, the study of Avicenna's *Canon on Medicine*.

> So I started [the study] of the principles of the *Canon* with my paternal uncle, the king of scientists ... Kamāl al-Dīn Abū al-Khayr Ibn al-Muṣliḥ-i Kāzerūnī, and with ... Shams al-Dīn Muḥammad Ibn AḤmad al-Ḥakīm al-Kīshī, then with the savant of the age, Sharaf al-Dīn Zakī al-Būshkānī, since they were famous for the teaching of this work and the distinguishing of the chaff from the grain, while having a clear view to the solution of its problems and the uncovering of its complexities. May the Lord bless them .... Yet, by virtue of this book being the most difficult composed in this art as far as comprehension, and the most straitened in terms of its course – this due to the inclusion of sagacious remarks, exact scientific formulations and wondrous points and extraordinary mysteries – the minds of the contemporaries were perplexed and the strength of others of the moderns failed to reach the apogees of their [course]. For the ideas included therein are the limits of the viewpoints of the foremost of the ancients and the extreme thoughts of the moderns, so that not one of them was capable of treating the book as it should be treated and therefore I despaired of them and likewise of the commentaries that I had encountered.[9]

At this point Shīrāzī lists some of the exceptional commentaries that he had come across and faults them for not adding to the discussion as it appears in Avicenna's book, but rather of "speaking on the topics that he had spoken" and "keeping silent about that which he had been silent." Despondent over the state of the commentaries on Avicenna's *Canon* Shīrāzī then sets out to meet the illustrious savant Ṭūsī:

> [So] I turned my attention to that city of knowledge and that face of the *ka'ba* of wisdom; the high, precious, holy, splendid presence and the elevated, immaculate, masterly and philosophical threshold ... of [Naṣīr al-Dīn Ṭūsī] may the Lord sanctify his soul and bless his tomb, [so that] some of the obscure points were clarified with others remaining obscure, since a mastery of the principles of theory is not sufficient for the comprehension of this book. Rather it is necessary, in addition, for the person to be a practiced physician with [experience] in the principles of treatment via the equilibration of the complexions.[10]

It is worth noting that Ṭūsī was by this point in the service of Hülegü, and Shīrāzī's tutelage under him would of necessity have been at Marāgha. Shīrāzī tells us that the subsequent stage of his project with respect to the *Canon* was to embark on an extended journey and to thus cast his net farther and wider for information pertaining to the *Canon*.

---

[8] Shīrāzī, *Durrat al-tāj li-ghurrat al-dabāj*, d; Minovi, "Mulla Qutb Shīrāzī," 166.

[9] Shīrāzī, *al-Tuḥfa al-sa'dīya fī al-ṭibb*, Suleimaniya MS 3649, 3r. Note this portion of the *Durrat al-tāj* edition that was used appears to have errors, and the Suleimaniya MS 3649 was used instead.

[10] Shīrāzī, *Durrat al-tāj li-ghurrat al-dabāj*, d.

> I then travelled to Khurāsān and from there to the cities of the 'Irāq-i 'ajam then to 'Irāq-i 'arab, Baghdād and its environs and from there to Rūm and I engaged in discussions with the scientists of these realms and the physicians of these parts and I asked them of the truths of these difficulties, and I benefited from what they possessed as far as detailed knowledge so that I had amassed what no one had amassed as far as [knowledge] and yet despite all of this effort and peregrinations even to Rūm, what was [unknown] in the book remained more than what was apparent.[11]

The subsequent episode that Shīrāzī includes in his autobiography is his service as Tegüder Aḥmad's ambassador to the Mamluk court, in 681 A.H. In his decades-long zeal for unlocking the mysteries of the *Canon*, Shīrāzī was apparently able to benefit from this diplomatic mission by obtaining new commentaries for the *Canon* in Cairo. At long last these manuscripts enabled Shīrāzī to embark on authoring his own commentary of the *Canon*:

> There I succeeded in obtaining three of the comprehensive commentaries on the *kullīyāt*: one from the ... philosopher 'Alā al-Dīn Abu al-Ḥasan 'Alī Ibn Abū al-Ḥazm al-Qurashī who is known as Ibn al-Nafīs, and the second from the [master] physician Ya'qūb Ibn Isḥaq al-Sāmerī and the third the physician Abū al-Faraj Ya'qūb Isḥaq al-Masīḥī known as Ibn al-Qiff and I succeeded in obtaining [as well] the responses of al-Sāmerī to the questions of the physician Najm al-Miftāḥ on some of the viewpoints of the book, [obtaining as well] a recension of the *Canon* by Hibbatallah Ibn Jamī' al-Yahūdī al-Maṣrī in which he refuted the Sheikh [i.e., Avicenna], and in addition some of the ... notes written by Amīn al-Daula ibn Tilmīdh upon the margins of the book, obtaining as well the book of ... the Imām 'Abd al-Laṭīf Ibn Yūsuf Ibn Muḥammad al-Baghdādī in which he refuted Ibn Jamī' [in regard to his recension of the Canon]. When I studied these commentaries and others which I had obtained, the remainder of the book became clear such that there did not remain within it obscurity or difficulty nor was there left room for disputation. And since I had collected what no person had collected in regard to the knowledge of the decipherment of this book and of the separation of what within it is as chaff to the grain I finally saw fit to write a commentary upon it so as to reduce the difficulty of the words, and to remove from the face of the meanings the mask of obscurity ... and [to provide] an indication of the responses to that which every commentator had [found objectionable, following a spirit of fairness and avoiding injustice and lack of due consideration] for to God we return and He is most worthy of our fear.[12]

Shīrāzī states that he started the composition for this work in 682 A.H. (i.e., 1283–1284 C.E.) and he also states "I gathered in it all that was unusual and difficult for others to collect, in as much as my intellect and my abilities permitted." He adds that his book was an "expansive commentary [based upon principles] that contained a multitude of questions and answers and lengthy marginalia and follow-up comments," and that it gained wide renown.[13] Indeed, the success of Shīrāzī's original commentary on the principles of the *Canon* was great enough (he claims) that he was approached and asked repeatedly to complete his commentary (perhaps for the remaining portions of the *Canon*). Among the reasons that Shīrāzī provided for refusing these requests were the perverse "constancy of the Fates" that forced

---

[11] Shīrāzī, *Durrat al-tāj li-ghurrat al-dabāj*, d.

[12] Shīrāzī, *Durrat al-tāj li-ghurrat al-dabāj*, d.

[13] Shīrāzī, *Durrat al-tāj li-ghurrat al-dabāj*, d.

him to leave his homeland on dangerous journeys, all the while preventing him from writing.[14] Also responsible were:

> a continuous string of cataclysms afflicting learned men [one following the other] until they had effaced the worksites of religion and until the pillars of religious law had weakened utterly, oppressing knowledge and its [practitioners] and obstructing from all directions its [valued offerings, so that] its minaret lay in ruins and all traces of it were obliterated.[15]

It is interesting to note that one of Shīrāzī's concerns in regard to the detrimental effect of the mayhem let loose by the Fates, was its effect on his acumen and judgement: "Some learned men do not issue fatwas on Saturday and Wednesday and claim [as their excuse] that holidays on Friday and Tuesday weaken understanding ... and if holidays are a single day ... so what then would you think of a 20 year long hiatus, without debates, study, [scholarly] work, and disputation."[16] (We should note here that the period 1281–1285 C.E. appears to have been particularly productive with respect to publications: In addition to the first edition of his commentary on the *Canon,* Shīrāzī's three works on *hay'a* belong to this period, as we will see). Shīrāzī's reference to the "string of cataclysms" is clarified somewhat in his description of how the dismal state of affairs finally comes to an end and is reversed:

> Until the Lord brought forth from it [i.e., religion] victory and triumph and provided the Muslims with strength and power, and the star of Islam appeared and the government of [Ghāzān] rose as the Sun upon the sleepers, may his elevated threshold be ever encompassed by the swords of victory....[17]

The cataclysms are then dated to the period subsequent to Shīrāzī's trip to Cairo (in 1282 C.E.) and the accession of Ghāzān in 1295 C.E. Presumably the death of Shīrāzī's patron Shams al-Dīn in 1284 C.E. was among the earliest of the cataclysms that Shīrāzī alludes to. In the remainder of the introduction to his commentary on the *Canon* Shīrāzī dedicates the work to his patron, the minister Sa‘d al-Dīn Sāvajī (d. 1311–1312 C.E.), and describes some of the details of what appears to be, in effect, a new edition of his work of 682 A.H. (i.e., 1283–1284 C.E.).[18]

3.3 Biographical Information in *Majma‘ al-ādāb*

As the librarian of the observatory at Marāgha Ibn al-Fuwaṭī apparently knew Shīrāzī personally (see, for example, Shīrāzī's *ijāza* in Sect. 3.5 of this chapter). Sadly, Ibn al-Fuwaṭī's original work has been lost, and what has survived is merely

---

[14] Shīrāzī, *Durrat al-tāj li-ghurrat al-dabāj*, dh.

[15] Shīrāzī, *Durrat al-tāj li-ghurrat al-dabāj*, dh.

[16] Shīrāzī, *Durrat al-tāj li-ghurrat al-dabāj*, dh.

[17] Shīrāzī, *Durrat al-tāj li-ghurrat al-dabāj*, dh.

[18] For more information on Shīrāzī's patron Sa‘d al-Dīn Sāvajī see Walbridge, *The Science of Mystic Lights*, 186. In his dedication Shīrāzī invokes the name of the ruler Ghāzān, as well.

an abridgment of the original. This is especially unfortunate because Ibn al-Fuwaṭī begins his biography of Shīrāzī by describing him as: "A learned man, who, were I to commence in describing I would [in so doing] require an entire volume by itself." As it is, the surviving text by Ibn al-Fuwaṭī only touches on two of the main episodes of Shīrāzī's life. The first is his trip to Marāgha seeking Ṭūsī's tutelage, for which Ibn al-Fuwaṭī supplies the date 658 A.H. (i.e., 1259–1560 C.E.). In describing this trip to Marāgha Ibn al-Fuwaṭī lists two of Shīrāzī's other teachers, as well:

> He arrived in Marāgha in the presence of our guide and master Naṣīr al-Dīn Ṭūsī in the year 658 and studied mathematics with him and studied with Najm al-Dīn al-Kātibī al-Qazwīnī that which he had composed on logic and with Mu'ayyad al-Dīn al-'Urḍī that which he had composed in astronomy and geometry and he wrote with his fine and comely hand all that he had studied and had achieved and [he exerted himself in his studies] night and day.[19]

In addition to Ṭūsī and al-'Urḍī, Najm al-Dīn al-Kātibī (d. 657 A.H./1276 C.E.) was one of the important scientists working at Marāgha.[20] He is one of four astronomers whose contributions are acknowledged by name in the planetary table compiled at Marāgha, the *Zīj-i Ilkhānī*.[21] He appears also to have served as Shīrāzī's supervisor a bit later in the young man's career, as we will see in Sect. 3.5 of this chapter.

The second episode that is captured in Ibn al-Fuwaṭī's surviving text is one on which Shīrāzī is silent, i.e., the episode involving his appointment as judge in Sīvās: "and he was appointed judge in Rūm and lived in Sīvās for a while then returned to Azarbāijān and became a resident of Tabriz."[22] As we will see this appointment would have preceded Shīrāzī's role as ambassador to Cairo. In his opening Ibn al-Fuwaṭī describes Shīrāzī as possessing "a prophetic disposition, divine knowledge, a noble soul, a towering mind, generosity and beneficence."[23] He concludes by noting that, upon his return from Cairo, Shīrāzī "busied himself with writing and research and his presence became the gathering place for the wise and learned men. And he was mild-tempered and witty in discussions. He was also intimate with sultans and viziers. He was born in 630 A.H. and he died in Tabriz in the year 710 A.H. and was buried in the Jarandāb [cemetery]."[24]

[19] Ibn al-Fuwaṭī, *Majma' al-adab fi mu'jam al-alqāb*, vol. 3, 440.

[20] M. Mohaghegh, "al- Kātibī, Nadjm al- Dīn Abu'l-Ḥasan 'Alī b. 'Umar," in *Encyclopaedia of Islam, Second Edition* (Brill Online, 2010), http://www.brillonline.nl/subscriber/entry?entry=islam_SIM-4023; Bar Hebraeus, *Tārīkh mukhtaṣar al-duwal* (Bayrūt: al-Maṭba'ah al-Kāthūlīkīyah lil-Ābā' al-Yasū'īyīn, 1890), 151; Walbridge, *The Science of Mystic Lights*, 11; Mudarris Razavi, *Aḥwāl wa Athār-i Muḥammad Ibn Muḥammad Ibn al-Ḥasan al-Ṭūsī*, 130.

[21] Mudarris Razavi, *Aḥwāl wa Athār-i Muḥammad Ibn Muḥammad Ibn al-Ḥasan al-Ṭūsī*, 130.

[22] Ibn al-Fuwaṭī, *Majma' al-ādāb fi mu'jam al-alqāb*, vol. 3, 441.

[23] Ibn al-Fuwaṭī, *Majma' al-ādāb fi mu'jam al-alqāb*, vol. 3, 441.

[24] Ibn al-Fuwaṭī, *Majma' al-ādāb fi mu'jam al-alqāb*, vol. 3, 441.

3.4 Biographical Information in *Tārīkh al-Islām*

Al-Dhahabī's biography of Shīrāzī as it appears in his monumental *Tārīkh al-Islām*, provides many additional details in regard to Shīrāzī's life. Al-Dhahabī was a generation younger than Ibn al-Fuwaṭī and though he doesn't mention his sources on Shīrāzī it is likely that some of the information in Ibn al-Fuwaṭī's *Majma' al-ādāb* (lost, as we have said) would have found its way into the *Tārīkh al-Islām*. The information that Shīrāzī himself cites in regard to his early schooling appears in al-Dhahabī, as well. This information is rather garbled, however, at least in the modern edition of al-Dhahabī's work: "He was born in Shīrāz in 634 A.H., his father was a doctor and his paternal uncle was of the learned men so he studied with them and with al-Shams al-Kutubī and with Sharaf al-Dīn Zakī and Zakī al-Barshakānī."[25] While al-Dhahabī correctly lists both Shīrāzī's father and uncle as his teachers, it is clear that by al-Kutubī he is referring to Shams al-Dīn al-Kīshī, and that Sharaf al-Dīn Zakī and Zakī al-Barshakānī both refer to the same person, i.e., Sharaf al-Dīn al-Būshkānī.[26]

Al-Dhahabī also lists 'Alā' al-Dīn Muḥammad Ibn Abu Bakr al-Ṭā'ūsi as having taught *fiqh*, or jurisprudence, to Shīrāzī, though this episode is apparently of a later period, when Shīrāzī had left Shiraz and was in Qazwīn (Qazvin).[27] Minovi writes of a meeting between Shīrāzī and a certain Ḍiā' al-Dīn Ṭūsī in Qazwīn. Shīrāzī relates the reason for his residence to Ḍiā al-Dīn who reports it in turn: "He said I was engaged in the practice of medicine, but I left the practice and started traveling and learned theology (*'ilm al-kalām*) and the other intelligible sciences (*al-ma'qūlāt*), but I was ever yearning and my soul would not be content. Yet, I had no knowledge of the transmitted sciences (*al-manqūlāt*) and especially of jurisprudence (*fiqh*). It is for this reason that I study with Sheikh 'Alā' al-Dīn."[28]

At this point al-Dhahabī briefly states Shīrāzī's early career as a young physician and his trip to Marāgha: "and he was made a physician in the hospital while he was young, and he travelled to Naṣīr al-Dīn al-Ṭūsī and joined his retinue and studied with him his commentary on *al-Ishārāt* and mathematics and *hay'a* and he excelled in these."[29] Of particular interest to our discussion is the fact that al-Dhahabī lists Ṭūsī as having taught *hay'a* to the young Shīrāzī (recall that in his autobiographical material Shīrāzī's stated purpose for seeking Ṭūsī was his desire to

---

[25] Dhahabī, *Tārīkh al-Islām wa-wafāyāt al-mashāhīr wa al-a'lām*, vol. 54, 101.

[26] See Sect. 3.2 of this chapter, and Walbridge, *The Science of Mystic Lights*, 9.

[27] See Walbridge, *The Science of Mystic Lights*, 12.

[28] Minovi, "Mulla Qutb Shīrāzī," 169. Walbridge identifies Ḍiā' al-Dīn as a grandson of Naṣīr al-Dīn Ṭūsī. See Walbridge, *The Science of Mystic Lights*, 12, and Mudarris Razavi, *Aḥwāl wa Athār-i Muḥammad Ibn Muḥammad Ibn al-Ḥasan al-Ṭūsī*, 40–42.

[29] Dhahabī, *Tārīkh al-Islām wa-wafāyāt al-mashāhīr wa al-a'lām*, vol. 54, 101; Mudarris Razavi, *Aḥwāl wa āthār-i Muḥammad Ibn Muḥammad Ibn al-Ḥasan al-Ṭūsī*, 246.

acquire medical knowledge). The commentary in question here is the one that Ṭūsī wrote on Avicenna's *al-Ishārāt wa al-tanbihāt*, or "Remarks and Admonitions."[30]

A considerable amount has been written about a purported antipathy between Shīrāzī and his teacher Ṭūsī. The origins of these accounts are generally the later historical sources. Some of this material has been refuted by Razavi.[31] As we saw Shīrāzī addresses his deceased teacher with the utmost respect in the *al-Tuḥfa al-sa'dīya*. Furthermore, his last book on *hay'a*, *Fa'altu fa lā talum*, is dedicated to Ṭūsī's son Aṣīl al-Dīn. Al-Dhahabī's history, however, includes a short comment that is quoted by later historians and that may have served as the source for the other, more fanciful, accounts. He says: "[Shīrāzī consorted] with Hulākū and Abaqā and he [i.e., Abaqa] said to him: 'you are the best student of [Ṭūsī] and he has grown old. Strive, therefore, so that you do not [miss] any of his knowledge. He replied: I have done so, and there does not remain for me a need [for it].'"[32] Given apparent esteem in which Shīrāzī held Ṭūsī, it is surprising to hear this rather strident claim of self-sufficiency, rather than a more modest confession of inadequacy. Indeed, it is not clear what to make of this strange (purported) remark.

Though it is impossible to recreate the circumstances in which this interview took place the account resonates dimly with an episode we saw reported in Rashīd al-Dīn's history on the eve of the siege of Baghdād. There Hülegü had reacted to advice by his astrologer as to the folly of attacking the Abassid capital by asking Ṭūsī's opinion. Rashīd al-Dīn says that Ṭūsī was "alarmed, as though this was a test," and quickly offered his full support for the siege of Baghdād.[33] Whatever the original conversation between Shīrāzī and Abaqa– the dim echo of which has reached us through the span of some 700 years – it would likely be a mistake, therefore, to think of Abaqā's comment as an innocent or benevolent remark. Ṭūsī, Shīrāzī, and their cohorts were of great strategic importance to the Ilkhans and the desire that Ṭūsī's work continue after his death would have been anything but a frivolous concern.[34] It is possible that at his audience with Abaqā, Shīrāzī, too, felt as though he was being tested, and was compelled therefore to give a short and expeditious answer.

The subsequent portion of al-Dhahabī's biography deals with Shīrāzī's judgeship in Anatolia and his mission to Cairo: "He then went to Rūm and the *Barvānāh* honored him and appointed him as the judge of Sīvās and Malaṭīya. And he went to Syria as the ambassador of [Tegüder] Aḥmad and when Aḥmad was murdered [Shīrāzī went back to court] and Arghūn honored him."[35] The *Barvānāh* or, more properly, Parvāne, in question is Mu'īn al-Dīn, the administrator appointed by the

30Mudarris Razavi, *Aḥwāl wa Athār-i Muḥammad Ibn Muḥammad Ibn al-Ḥasan al-Ṭūsī*, 246.

31Mudarris Razavi, *Aḥwāl wa Athār-i Muḥammad Ibn Muḥammad Ibn al-Ḥasan al-Ṭūsī*, 71.

32Dhahabī, *Tārīkh al-Islām wa-wafāyāt al-mashāhīr wa al-a'lām*, vol. 54, 101.

33Rashīd al-Dīn Ṭabīb, *Jāmi' al-tawārīkh*, 717.

34Razavi quotes an uncited source as to the fact that Ṭūsī bequeathed his work on the *Zīj-i Ilkhānī* to his son Aṣīl al-Dīn and to Shīrāzī; *Aḥwāl wa Athār-i Muḥammad Ibn Muḥammad Ibn al-Ḥasan al-Ṭūsī*, 32.

35Dhahabī, *Tārīkh al-Islām wa-wafāyāt al-mashāhīr wa al-a'lām*, vol. 54, 101.

Mongols for Anatolia on the eve of Hülegü's campaigns in Persia.[36] As we saw in Chap. 2, Mu'īn al-Dīn payed with his life in 1277 C.E., for allegedly intriguing with the Mamluk ruler Baybars.[37] If, therefore, al-Dhahabī is correct in claiming that Shīrāzī's residence in Anatolia was at the behest of the Parvāne, then this would date Shīrāzī's appointment as judge to the period prior to 1277 C.E. which is the date of the Parvāne's execution, and probably before 1275 C.E. which is the date for the commencement of Baybars's adventure in Anatolia.[38] It should be noted here that Shīrāzī's translation into Persian of Ṭūsī's *Taḥrīr-i Uqlīdus* (*Exposition of Euclid*) is dedicated to this statesman.[39]

At some point after his return from Cairo, though al-Dhahabī does not make clear exactly when, Shīrāzī appears to have settled in Tabriz and focused on the study of ḥadīth literature. As Wiedemann suggests this could very well refer to the end of Shīrāzī's life.[40] That this may have been the case is supported by al-Dhahabī's narrative which switches here from an episodic format to a list of general remarks. In particular, al-Dhahabī lists here four of Shīrāzī's works: "and he is the author of books, among them the *Ghurrat al-Tāj* [sic] on philosophy and a commentary on *al-Asrār* [sic] by the murdered al-Suhrawardī, and a commentary on the *kullīyāt* and a commentary on *al-Mukhtaṣar* by Ibn al-Hājib."[41] Of these *Ghurrat al-Tāj* is clearly the *Durra*. Suhrawardī's work is the *Sharḥ ḥikmat al-ishrāq*, which was described briefly in Chap. 1. The *kullīyāt* in question here can only be the book of Avicenna (on which Shīrāzī wrote a commentary in the introduction of which he included his autobiography). *Al-Mukhtaṣar* appears to refer to the abridgment by Ibn Hājib of his own *Muntahā al-su'āl wa al-āmāl fī 'ilmay al-uṣūl wa al-jadal*.[42]

The remainder of al-Dhahabī's article describes the personal characteristics of Shīrāzī, noting especially his intellectual brilliance, his generosity, and his piety, but noting as well Shīrāzī's irreverence, his ability to play the *rubāb*, his fondness for

---

[36]Cahen, *Pre-Ottoman Turkey*, 273–276.

[37]Cahen, *Pre-Ottoman Turkey*, 276–291.

[38]Cahen, *Pre-Ottoman Turkey*, 286.

[39]Mir, *Sharḥ-i ḥāl wa asār-i 'allāmah Qutb al-Dīn Mahmud Ibn Mas'ud Shīrāzī, danishmand-i 'ali qadr-i qarn-i haftum, (634–710 A.H.)*, 69. A firm date for this work should be particularly useful in understanding the period in question.

[40]Dhahabī, *Tārīkh al-Islām wa-wafāyāt al-mashāhīr wa al-a'lām*, vol. 54, 101; E. Wiedemann, "Ḳuṭb al- Dīn Shīrāzī, Maḥmūd b. Mas'ūd b. Muṣliḥ," in *Encyclopaedia of Islam, Second Edition*, Edited by: P. Bearman. (Brill Online, 2010), http://www.brillonline.nl/subscriber/entry?entry=islam_SIM-4581. Shīrāzī relates here that he studied the "Commentary on the sunna" with a certain Muḥyi al-Dīn. I have not been able to locate additional information about this figure, but he is likely the same Muḥyi al-Dīn that is referenced in the *ijāza* that appears at the beginning of al-Sallāmī's discussion; see Sect. 3.4, of this chapter.

[41]Dhahabī, *Tārīkh al-Islām wa-wafāyāt al-mashāhīr wa al-a'lām*, vol. 54, 101.

[42]H. Fleisch, "Ibn al- Ḥādjib, Djamāl al-Dīn Abū 'Amr 'Uthmān b. 'Umar b. Abī Bakr al-Mālikī," in *Encyclopaedia of Islam, Second Edition*, Edited by P. Bearman. (Brill Online, 2010), http://www.brillonline.nl/subscriber/entry?entry=islam_COM-0324; See also, Walbridge, *The Science of Mystic Lights*, 189.

wine (some of which would have been questionable behavior for a scholarly man of Shīrāzī's reputation). Al-Dhahabī also adds: "And in the end he continued to serve, teaching *al-Kashshāf, al-Qānūn, al-Shifā'*, and the ancient (*awā'il*) sciences. We ask the Lord, blessed and most high, for salvation."[43] Here, by *al-Kashshāf* al-Dhahabī is referring to *al-Kashshāf 'an ḥaqā'iq al-tanzīl*, "Unveiler of the Realities of Revelations," the renowned Qur'anic commentary by Zamakhsharī (1075–1144 C.E./467–538 A.H.).[44] *Al-Qānūn* is Avicenna's *Canon* to which we have made numerous references in this chapter. The mention of neither of these books, however, would have compelled al-Dhahabī to invoke the name of Allah. Instead it is presumably the last two of the items on his list that al-Dhahabī found alarming: *al-Shifā'*, "The Healing," by Avicenna, containing the author's Aristotelian and Neoplatonic philosophy. The ancient sciences were held as suspect in various eras by many scholars in the Islamic world; here, al-Dhahabī appears to betray his ambivalence about these branches of knowledge.[45]

Al-Dhahabī concludes his article on Shīrāzī by stating:

> And God knows his intentions for of what was apparent we have spoken and what was hidden was finer [still, no doubt]. And he possessed excellent qualities, virtue, and upstanding morals. May the Lord forgive his sins and ours. Amen! For he was a sea of knowledge and a possessor of acumen, and mathematics was his most excellent field. I have witnessed his students honor him greatly.[46]

3.5 Biographical Information in *Tārīkh 'ulamā' Baghdād*

Prior to embarking on Shīrāzī's biography proper, the published version of al-Sallāmī's article partially reproduces an *ijāza*, or license, that Shīrāzī purportedly wrote for Ibn al-Fuwaṭī.[47] In this *ijāza*, Shīrāzī grants the licensee the permission to transmit two works: *Sharḥ al-sunna*, a commentary on the prophetic tradition,

---

[43] Dhahabī, *Tārīkh al-Islām wa-wafāyāt al-mashāhīr wa al-a'lām*, vols. 54, 101.

[44] C. Versteegh, "al- Zamakhsharī, Abu'l- Ḳāsim Maḥmūd b.'Umar," in *Encyclopaedia of Islam*, Second Edition (Brill Online, 2010), http://www.brillonline.nl/subscriber/entry?entry=islam_SIM-8108; W. Madelung, "al- Zamakhsharī, Abu 'l- Ḳāsim Maḥmūd b. 'Umar (Contributions in the fields of theology, exegesis, ḥadīth and adab).," in *Encyclopaedia of Islam, Second Edition* (Brill Online, 2010), http://www.brillonline.nl/subscriber/entry?entry=islam_COM-1469. See also, Walbridge, *The Science of Mystic Lights*, 188.

[45] For a discussion of the tension between the religious sciences and the Ancient sciences in Shīrāzī's era see, Robert Morrison, *Islam and Science, The Intellectual Career of Niẓām al-Dīn al-Nīsāburī* (London: Routledge, 2007).

[46] Dhahabī, *Tārīkh al-Islām wa-wafāyāt al-mashāhīr wa al-a'lām,* vol. 54, 102.

[47] Sallāmī, *Tārīkh 'ulamā' Baghdād al-musammá muntakhab al-mukhtār*, 176; G. Vajda, "Idjāza," in *Encyclopaedia of Islam, Second Edition*, Edited by P. Bearman. (Brill Online, 2010), http://www.brillonline.nl/subscriber/entry?entry=islam_SIM-3485.

by Ḥusayn Ibn Mas'ūd al-Baghawī (d. 1122 C.E./516 A.H.),[48] and *Jāmi' al-uṣūl*, by Majd al-Dīn Abū al-Sa'ādat al-Mubārak Ibn al-Athīr (1149–1210 C.E./544–606 A.H.).[49] In the conclusion of the *ijāza* Shīrāzī states:

> I hereby grant permission for the transmittal of these two books on my authority and likewise other texts, audited recited ... or related,[50] subject to the conditions stipulated by the people of transmission [i.e., the customary stipulations]. I am to be held inculpable insofar as modifications, distortions, transformations, and scribal errors. I ask the Lord to extend the longevity of the licensees through knowledge, so that submerged in it they may discover its treasures and so that the sea-shells of knowledge may yield their riches to them. May the Lord grant them success in the goodly action that is ... that point upon which the wayfarers to the limits of virtue affix their gaze.[51]

Al-Sallāmī then starts off the biography of Shīrāzī by rendering his life up to shortly before Ṭūsī's death as we have seen it before, with some minor modifications:

> He worked under his father and his paternal uncle and under al-Shams al-Kutubī [sic] and Zakī al-Barsakānī [sic]. And when his father died he was 14 years old and he was appointed to his father's position in the Muẓaffarī hospital in Shiraz, then he travelled when he was twenty something, heading for Naṣīr al-Dīn and accompanied him and studied his philosophical works and *hay'a* and he excelled in these [so that Ṭūsī would call him] the "pole of the sphere of existence" and he travelled with him to Khurāsān and then he returned to Baghdād and lived in the Niẓāmīya and the Ṣāḥib Dīwān [i.e., Shams al-Dīn Juwaynī] honored him and he consorted with Hülegü and Abaqā and Abaqā said to him "you are his best student," pointing to Ṭūsī "and he is approaching death, so strive so that you do not miss anything of his knowledge." He replied, "I have done so and no longer have I need for additional [knowledge]."[52]

That the erroneous rendition of Shīrāzī's early teachers is similar to al-Dhahabī's this is not surprising as al-Sallāmī expressly cites al-Dhahabī and Ibn al-Fuwaṭī as his sources for Shīrāzī's biography.[53] The information seen here that is missing in al-Dhahabī (and the likely source of which must therefore be Ibn al-Fuwaṭī's lost work) is Ṭūsī's characterization of the young Shīrāzī, which contains a pun on Quṭb al-Dīn's name; *quṭb* being the word for pole in Arabic. This speaks of Ṭūsī's affection and esteem, and may explain, as well, the source for Qutb al-Dīn's title. In addition, the episode of the trip to Khurāsān that is described by al-Sallāmī is

---

[48]J. Robson, "al- Baghawī, Abū Muḥammad al-Ḥusayn b. Mas'ūd b. Muḥ. al-Farrā' (or Ibn al-Farrā')," in *Encyclopaedia of Islam, Second Edition*, Edited by: P. Bearman. (Brill Online, 2010), http://www.brillonline.nl/subscriber/entry?entry=islam_SIM-1024.

[49]F. Rosenthal, "Ibn al- Athīr," in *Encyclopaedia of Islam, Second Edition* (Brill Online, 2010), http://www.brillonline.nl/subscriber/entry?entry=islam_SIM-3094.

[50]In the list of works the transmission of which was sanctioned by his *ijāza,* Shīrāzī includes the category al-mustajāzāt, suggesting a set of works for which an ijāza was solicited by the author himself.

[51]Sallāmī, *Tārīkh 'ulamā' Baghdād al-musammá muntakhab al-mukhtār*, 176.

[52]Sallāmī, *Tārīkh 'ulamā' Baghdād al-musammá muntakhab al-mukhtār*, 177.

[53]Sallāmī, *Tārīkh 'ulamā' Baghdād al-musammá muntakhab al-mukhtār*, 177–179.

not mentioned by al-Dhahabī (nor does it appear in the surviving portion of Ibn al-Fuwaṭī's chronicle). It is almost certainly, however, the same trip that Shīrāzī mentions in his autobiography (while perplexingly omitting the fact he undertook this trip as a member of Ṭūsī's party). The account of Shīrāzī's stay at the Niẓāmīya in Baghdād, and the patronage of Shams al-Dīn Juwaynī appear to be the earliest surviving description we have of this portion of Shīrāzī's life.

In al-Sallāmī's rendition of the exchange between Shīrāzī and Abaqā the new detail is the presence of Ṭūsī, himself, and the fact that Ṭūsī is nearing his death. Though contriving a scenario for this exchange would be purely conjectural, the fact that in al-Sallāmī's account we see Ṭūsī as quite apparently failing, should at least allow the possibility that Shīrāzī's shortness in responding to Abaqā's injunction could have been driven by his desire to avoid tormenting his teacher by the unwanted attention of a sober-minded and pragmatic Ilkhanid ruler. That the meeting between Shīrāzī and Abaqā took place near the end of Ṭūsī's life is also supported by the fact that during Ṭūsī's last visit to Baghdād (to which he had gone for the sake of attending to *awqāf* accounts, and in which he died) he appears to have been accompanied by Abaqā,[54] with at least one account depicting both Shīrāzī and Abaqā as present at Ṭūsī's deathbed.[55]

The account of Shīrāzī's relocation to Anatolia appears in al-Sallāmī – together with what is the only surviving text that mentions anything about Shīrāzī's children, as follows:

> So he went to Rūm and "The Eagle" honored him and ... appointed him as judge of Sīvās and Malaṭīya and [so] he went with his children to Rūm. And Ibn al-Fuwaṭī relates that he was always [deep] in thought and engaged in writing and his hand was never [devoid] of a pen. And people would gather to him and [benefit from his company]. And he was good-humored and witty and generous.[56]

Based on the parallel account in al-Dhahabī, the character referred to as "The Eagle" is perhaps Mu'īn al-Dīn (i.e., the Parvāne) himself, though I have not found another reference to him by this name.[57]

Another episode for which there is no surviving account prior to its appearance in al-Sallāmī's work is Shīrāzī's residence in Juwayn (Jovayn), the home district of Shams al-Dīn and 'Ala' al-Dīn: "And he left Azarbāijān and resided for a spell in the school which Shams al-Dīn Muḥammad Juwaynī had built in Juwayn – the [responsibilities of it teaching program] that he had conferred upon Najm al-Dīn al-Kātībī al-Qazwīnī. And Qutb al-Dīn was the assistant in his teaching."[58] Recall that according to the surviving biography of Ibn al-Fuwaṭī al-Kātibī was Shīrāzī's teacher of logic at Marāgha (see Sect. 3.3 of this chapter). The dates for this episode are unknown. What can be said with reasonable certainty is that it was before

[54]Mudarris Razavi, *Aḥwāl wa āthār-i Muḥammad Ibn Muḥammad Ibn al-Ḥasan al-Ṭūsī*, 35.

[55]Mudarris Razavi, *Aḥwāl wa āthār-i Muḥammad Ibn Muḥammad Ibn al-Ḥasan al-Ṭūsī*, 35.

[56]Sallāmī, *Tārīkh 'ulamā' Baghdād al-musammá muntakhab al-mukhtār*, 177.

[57]It should be noted that in Arabic "Eagle" and "Vulture" are designated by the same word, *al-nasr*.

[58]Sallāmī, *Tārīkh 'ulamā' Baghdād al-musammá muntakhab al-mukhtār*, 178.

Shīrāzī's residence in Anatolia. Shīrāzī himself tell us that by 1274 C.E. he was in Konya (Persian/Arabic Qūnīya) studying ḥadīth and other topics with Ṣadr al-Dīn Qūnawī.[59] In this case the period between 1269 C.E. (the end-date of Shīrāzī's trip to Khurāsān with Ṭūsī) and 1274 C.E. would have seen Shīrāzī in Juwayn serving as assistant to Kātibī, as well as in Baghdād at the Niẓāmīya. The dates of Shīrāzī's study with Ṭa'ūsī in Qazwīn are not known, but since it is hardly conceivable that he would have done this after his appointment as judge by Mu'īn al-Dīn (if we are to believe al-Dhahabī), then Shīrāzī's Qawvīn episode and the other two belong to the period of roughly 1269 C.E. to c. 1274 C.E. If Shīrāzī's ordering of events is assumed accurate this would mean that he spent the period prior to 1274 C.E. in Marāgha, Khurāsān, Qazwīn, and Baghdād, prior to traveling to Anatolia and settling in Konya. The appointment as judge in Sīvas would have been prior to 1277 C.E. and he may have remained in Sīvās (if not serving as judge continuously) until 1281 C.E. when he completed the *Nihāya*.

Al-Sallāmī's description of Shīrāzī's trip to Anatolia is unfortunately muddled, however, by the existence of a second account of what appears to be the same event. Immediately after the Juwayn episode al-Sallāmī has the following:

> And Shams al-Dīn appointed him as judge in Anatolia so he [went there] and took up residence in Sīvās and the seekers of knowledge enjoyed and benefited from his presence and he wrote [there] on the principles of *fiqh* and a commentary on Ibn al-Hājib's book and authored the *Ikhtīyārāt al-Muḍafarrīya* [sic] and the commentary on the *Miftāḥ* of Sakkākī and a commentary on the *kullīyāt* [of the Canon] by Avicenna and he wrote the book the *Tuḥfa* on the science of *hay'a* as well as other treatises and books.[60]

Though it is not clear what to make of the apparently conflicting accounts of how Shīrāzī was appointed as judge in Sīvās, it should be noted here that as a vassal state with what was effectively an Ilkhan-appointed viceroy in the person of Mu'īn al-Dīn, the Seljuks were ultimately under the control of the Mongol Ilkhans. That Shams al-Dīn alone was responsible for Shīrāzī's appointment and that he did this after the death of the Parvāne (i.e., sometime after 1277 C.E.), is within the realm of possibility, though this would render al-Dhahabī's account as completely wrong. Rather than dismiss out of hand al-Dhahabī's assertion that Mu'īn al-Dīn was responsible for Shīrāzī's appointment as judge, a more probable narrative would have both administrators, one belonging to the ruling state and one to the vassal, as having effected Shīrāzī's appointment in Sīvās. Melville includes a telling detail about the Seljuk monuments in Sīvās in his "Cambridge History of Turkey" article on Anatolia under Mongol rule: despite the fact that the Çifte Minare Medresesi (i.e., the "Madrasa of the Twin Minarets") was founded by Shams al-Dīn in 1272 C.E., the inscriptions on this monument do not include the names of either the Mongol or the Seljuk ruler. This fact emphasizes both Shams al-Dīn Juwaynī's personal

[59] Walbridge, *The Science of Mystic Lights*, 14.

[60] Sallāmī, *Tārīkh 'ulamā' Baghdād al-musammá muntakhab al-mukhtār*, 178.

interest in Sīvās, as well as the extent of his power and prestige there.[61] Given the contradictory accounts of Shīrāzī's appointment (as well as Shams al-Dīn's personal interest in Sīvās), the best we can do now is assume that both administrators – Shams al-Dīn from the ruling state and Mu'īn al-Dīn from the vassal state – were in some form involved in appointing Shīrāzī to judge in Sīvās, some time before 1277 C.E. (and, presumably, after his stay in Konya in 1274 C.E.).

Of the books listed earlier, several are known to have been completed while Shīrāzī was in Sīvās these are the *Nihāya*, the *Tuḥfa* and the earlier edition of the commentary on the *Canon*. According to Minovi Shīrāzī's commentary on Ibn al-Hājib was dedicated to Shams al-Dīn Juwaynī and so must predate this statesman's execution in December 1284 C.E.[62] As we will see in Chap. 4, the earliest historical evidence for Shīrāzī's residence in Sīvās is apparently the *Nihāya* itself, which was completed in November 1281 C.E. It is therefore certainly plausible that the commentary on Ibn al-Hājib's book was also written in Sīvās. According to Walbridge, the commentary on the *Key* [*to the sciences*] of Sakkākī, is dedicated to the dedicatee of the *Durra* and so likely belongs to the same period as this later work. If so, this commentary would have been completed long after Shīrāzī's return from Sīvās, since the *Durra* belongs to the last decade of Shīrāzī's life.[63]

In regard to Shīrāzī's embassy to Cairo al-Sallāmī states:

> He then returned to the presence of the Sultan Abaqā and when Sultan Aḥmad Takudār followed immediately in the footsteps of Abaqā he could not find anyone worthy of being sent to Egypt and Syria except for [Shīrāzī], who went accompanied by a letter in the year [6]81 A.H. to [Sultan Qalāwun] and he returned to Azarbāijān and we heard [!] the contents of the letter in his own words, with most of it having been composed by him. And when Maulānā Quṭb al-Dīn [i.e., Shīrāzī] came and delivered the message [of Qalāwūn?] to the Sultan [i.e., Aḥmad], he cast his walking staff to the ground in Tabriz [i.e., ending his journeys there]. [64]

Al-Sallāmī's account here is slightly more detailed than al-Dhahabī's in regard to Shīrāzī's whereabouts immediately prior to his ambassadorship to Cairo. The *Nihāya* was completed in November 1281 C.E./Sha'bān 680 A.H. with Shīrāzī in Sīvās (as we will see in Chap. 4). Rashīd al-Dīn reports that Abaqā died 4 months

---

[61]C. Melville, "Anatolia Under the Mongols," in *Byzantium to Turkey, 1071–1453*, vol. 1, The Cambridge History of Turkey (Cambridge: Cambridge University Press, 2009), 73. Despite the fact that the remains of this splendid monument were under heavy repairs in the summer of 2009 during my short visit to Sivas, the quality of the stone-carving and the tile-work were a clear indication of the rather astounding level of craftsmanship that had gone into its construction. The monument as it stood in the 13th century would have been opulent, indeed. One of the extant manuscript copies of the *Nihāya/Limit* was apparently written at this madrasa (see Chap. 4, note 1).

[62]This work was already listed in Sect. 3.3 of this chapter. See Minovi, "Mulla Qutb Shīrāzī," 195. See also, Walbridge, *The Science of Mystic Lights*, 189.

[63]Walbridge, *The Science of Mystic Lights*, 190; Minovi, "Mulla Qutb Shīrāzī," 190.

[64]Sallāmī, *Tārīkh 'ulamā' Baghdād al-musammá muntakhab al-mukhtār*, 178. It is clear that al-Sallāmī is quoting Ibn al-Fuwaṭī directly here, since he claims to have learned of the contents of the diplomatic letters from Shīrāzī himself.

later in Dhū al-Ḥijja. Tegüder's accession to the Ilkhanid throne did not happen until the 13th of Rabī' I of 681 (i.e., June 1282). If, therefore, al-Sallāmī is correct in his report, Shīrāzī moved from Sīvās to the court in Tabriz shortly after the completion of the *Nihāya* and stayed there for a little under a year before being sent to Cairo shortly after Tegüder's accession to power.

We do know that Shīrāzī was back in Sīvās by Jumādā I of 684 A.H., because this is the date for the *Tuḥfa*, which was completed in Sīvās. The period preceding the completion of the *Tuḥfa* would have been particularly strife-ridden as it saw the revolt of Arghūn and the ensuing death of Tegüder as well as the death of both Shams al-Dīn Juwaynī and his brother 'Alā' al-Dīn. Indeed, if we are to believe Shīrāzī's autobiography his mission to Cairo would have occurred shortly prior to the onset of what he termed "a string of calamities." As we saw in Sect. 3.2, Shīrāzī himself viewed the accession of Ghāzān as the end of a long and calamitous era. Certainly, Ghāzān's conversion to Islam would have been partially responsible for the praise that Shīrāzī bestows upon Ghāzān in his autobiography. Yet, the conversion per se can not have been the sole source of Shīrāzī's laudatory tone, as earlier Ilkhanid rulers with whom Shīrāzī was on good terms had been non-muslims. Indeed, statements in both al-Dhahabī and al-Sallāmī describing the great esteem that Ghāzān had for Shīrāzī may indicate a sort of restoration of Shīrāzī at the Ilkhanid court following a period of obscurity. We should recall, however, that Shīrāzī appears to have retained his importance even under Arghūn (as indicated by Rashīd al-Dīn's accounts), and that if he suffered any professional or public setbacks due to the unspecified cataclysms to which he alludes, these would have had to have occurred during the reign of Gaykhātū (1291–1295 C.E.), whose name along with that of short-reigned Bāydū (1295 C.E.), does not appear in any biographical texts related to Shīrāzī.

The remainder of al-Sallāmī's biography describes Shīrāzī's work habits, his piety, and his disregard for worldly things. As al-Sallāmī himself states much of this is taken from al-Dhahabī and Ibn al-Fuwaṭī. The new bits of information that appear in the remainder of al-Sallāmī's article may again have been taken from the lost work of Ibn al-Fuwaṭī. In regard to Shīrāzī's compositions al-Sallāmī states:

> And he was dedicated to composition and writing and study and he composed the book *Durrat al-Tāj* for the Malik Dūbāj the king of Gīlān. And he composed for Maulānā Aṣīl al-Dīn al-Ḥasan Ibn Naṣīr al-Dīn the book *Fa'altu fa lā talum* [i.e., *I have done it, so don't blame me*], which is a strange book in which he censures someone who didn't understand what he had said, [composing] as well as other works in the intelligible and transmissible arts.[65]

As we have stated previously Aṣīl al-Dīn was Naṣīr al-Dīn Ṭūsī's son and he was put in charge of the Marāgha observatory after the death of his father. Al-Sallāmī also states that Shīrāzī's students composed poems in his honor and that these were collected in a book.[66]

---

[65]Sallāmī, *Tārīkh 'ulamā' Baghdād al-musammá muntakhab al-mukhtār*, 178–179.

[66]Sallāmī, *Tārīkh 'ulamā' Baghdād al-musammá muntakhab al-mukhtār*, 179.

3.6 Biographical Information in *al-Durar al-Kāmina*

Al-'Asqalānī's short biography of Shīrāzī repeats the information we have seen in Ibn al-Fuwaṭī, al-Dhahabī and al-Sallāmī as far as Shīrāzī's intellectual prowess, his personal habits, his humility, and his sense of humor. When listing Shīrāzī's book he includes the title *Sharḥ al-ishrāq* (sic), referring to Shīrāzī's commentary on Suhrawardī. This is closer to the actual title of the work *Sharḥ ḥikmat al-ishrāq*, and it indicates that the reason for the error in al-Dhahabī's biography was a misreading of *asrār* for *ishrāq*.

In addition, al-'Asqalānī provides two bits of information that do not appear in the previous histories examined for this chapter. The first is his statement that the title by which Shīrāzī is known by the cognoscenti is *al-shāriḥ al-'alāma*, or "the Commentator Savant."[67] This title underscores Shīrāzī's great prestige as an intellectual. The second new piece of information by al-'Asqalānī is included in the following statement: "And when Ṣafī al-Dīn al-Muṭrib [i.e., Ṣafī al-Dīn the minstrel] went to him, he gave him two thousand dirhams, and he taught *al-Kashshāf*, the *Canon*, *al-Shifā'* and other books in Damascus."[68] The first part of the statement clearly parallels a statement in al-Dhahabī: "And when Ṣafī al-Dīn 'Abd al-Mu'min al-Maṭarī went to him he gave him two thousand dirhams." (Al-Maṭarī is an obvious misreading of al-Muṭrib.) The second part of the statement also appears to be derived from al-Dhahabī who, as we saw, said Shīrāzī taught *al-Kashshāf*, the *Canon, al-Shifā'* and the ancient sciences. In his version al-'Asqalānī has replaced "the ancient sciences" with *ghayrahā*, or "others." But he has also inserted the information about where this teaching supposedly took place, i.e., in Damascus. Walbridge repeats this information on al-'Asqalānī's authority.[69]

The problem with this additional bit of information, however, is that the only other records of Shīrāzī being in Damascus refer to the trip undertaken as a member of Sultan Tegüder Aḥmad's embassy. The Mamluk historian al-Ẓāhir who was a courtier in Cairo states emphatically, however, that the Sultan asked his deputies to make sure that "no one sees [the Ilkhan contingent] or associates with them, nor should anyone speak a word with them, and that they should travel only at night."[70] Al-Ẓāhir also states that on the return trip the same security measures were taken "so that no one associated with them, or glanced at them, or saw them ... and they reached Aleppo on the sixth of Shawwāl of the year 681, and [from there] made for their own lands."[71] Given all this, it is very difficult to imagine how Shīrāzī

[67] Ibn Ḥajar al-'Asqalānī, *al-Durar al-kāminah fī a'yān al-mi'ah al-thāminah*, vol. 5, 109.

[68] Ibn Ḥajar al-'Asqalānī, *al-Durar al-kāminah fī a'yān al-mi'ah al-thāminah*, vol. 5, 108.

[69] Walbridge, *The Science of Mystic Lights*, 17.

[70] Ibn 'Abd al-Ẓāhir, *Tashrīf al-ayyām wa-al-'uṣūr fī sīrat al-malik al-manṣūr*, pt. 2, 6.

[71] Ibn 'Abd al-Ẓāhir, *Tashrīf al-ayyām wa-al-'uṣūr fī sīrat al-malik al-manṣūr*, pt. 2, 16. Indeed given the great suspicion that existed between the two polities it is rather surprising that Shīrāzī was so successful in garnering his manuscripts of the *Canon*.

would have been allowed to lecture or to teach during this trip. Since al-'Asqalānī's statement is the only surviving reference to this teaching, his insertion of Damascus in the account that he appears to have gotten from al-Dhahabī is almost certainly in error.

3.7 Biographical Information from Shīrāzī's *Durra*

Shīrāzī states that he received the "khirqa in blessing" from his father. The bestowal of this woolen frock normally signifies ones status as a sufi or a disciple. Walbridge adds, however, that this *khirqa* was given "in blessing" implies that it was given as a sign of favor rather than a formal signifier of Shīrāzī having been inducted into sufism. The source for this biographical information is the *Durra*.[72] Elsewhere in the same work Shīrāzī describes receiving a *khirqa* as an adult: "The [humble] pauper who is the author if these words ... received the *khirqa* from the hands of the Sheikh Najīb al-Dīn 'Alī Ibn Buzghush al-Shīrāzī, may the Lord sanctify his soul, and he [in turn] received it from the sheikh of sheikhs Shahāb al-Dīn al-Suhrawardī, may the Lord rest his soul."[73]

Of the authors that we have seen earlier in the chapter, al-Dhahabī writes: "And he was one of the smartest men of the age, and was witty and sharp and did not carry concerns of the [impermanent] world with him. And he wore the garbs of the sufis."[74] Al-'Asqalānī writes: "And he consorted frequently and freely with kings, and was witty, and bright, and did not carry any concerns, and did not [ever] alter his sufi garb."[75] Al-Sallāmī does not include a reference to Shīrāzī's sufi garbs, but says instead: "he was not concerned with his clothes and he did not [claim the seat of honor] in gatherings."[76] It is reasonably clear from these words that Shīrāzī was a sufi (or at least a disciple of sufism) for all of his adult life. It is in view of this information that his somewhat unorthodox personal habits with respect to music, and alcohol, and his apparent disregard for worldly pomp should ultimately be understood.[77]

---

[72]Quṭb al-Dīn Shīrāzī, *Durrat al-tāj li-ghurrat al-dabāj*, ch; See also Walbridge, *The Science of Mystic Lights*, 9.

[73]Mir, *Sharḥ-i ḥal wa āsār-i 'allāmah Qutb al-Dīn Maḥmūd ibn Mas'ūd Shīrāzī, danishmand-i 'ali qadr-i qarn-i haftum, (634–710 A.H.)*, 19; Walbridge, *The Science of Mystic Lights*, 10.

[74]Dhahabī, *Tārīkh al-Islām wa-wafāyāt al-mashāhīr wa al-a'lām*, vol. 54, 101.

[75]Ibn Ḥajar al-'Asqalānī, *al-Durar al-kāminah fī a'yān al-mi'ah al-thāminah*, vol. 5, 108.

[76]Sallāmī, *Tārīkh 'ulamā' Baghdād al-musammá muntakhab al-mukhtār*, 179.

[77]Amitai, "Sufis and Shamans."

3.8 The Final Decades

Shīrāzī appears to have spent the last two decades of his life (i.e., c. 1290–1311 C.E.) in Tabrīz. Though information that can be traced to this period of Shīrāzī's life is scarce, all of the sources we have examined state or imply that he remained active in teaching, and several imply as well that he focused more on the religious sciences as time wore on. As we saw, Shīrāzī publicly affirmed his high regard for Ghāzān (1295–1304 C.E.) and of the sources we have seen al-Dhahabī and al-'Asqalānī also mention Ghāzān by name, implying that this esteem was mutual. In addition al-Dhahabī and al-'Asqalānī imply that Shīrāzī was able to intercede to Ghāzān on behalf of others. In a rather odd remark al-Dhahabī implies that Shīrāzī was intimate, as well, with Öljeitü, the subsequent Ilkhanid ruler, and the one during whose reign Shīrāzī died: "And [Shīrāzī] had mastered magical tricks and he played the rubāb and he presented variegated jests in the presence of *Kharband* and [also] in his lessons."[78] *Khar-bandeh* was one of the titles of Öljeitü and it is reasonably certain that al-Dhahabī is referencing this ruler. There is no record that Shīrāzī followed the court of Öljeitü when the ruler relocated from Tabrīz to Sulṭānīya c. 1305.[79] However, al-Dhahabī's remarks suggest that he was well-regarded at court to his death in 1311 C.E. The expenses for the lavish funeral appear to have been covered by 'Izz al-Dīn Ṭayyibī, an affluent disciple of Shīrāzī.[80]

As opposed to Shīrāzī's peregrinations in his youth, and adulthood this final phase of Shīrāzī's life appears to have been a relatively settled period. Walbridge places Shīrāzī in Gīlān c. 1305, based on the fact that he dedicates his *Durra* to the ruler "Amīra al-Dabāj" who was one of the rulers of Gīlān at the time.[81] It is rather difficult to believe that Shīrāzī would have undertaken this journey to the untamed region of Gīlān, during the last decade of his life. It is known that al-Dabāj paid a visit to the Ilkhanid court prior to Öljeitü's unsuccessful campaign in Gīlān c. 1306 C.E.[82] It is therefore much more likely that Shīrāzī completed his book in Tabrīz and dedicated it to the visiting dignitary from the frontier area, perhaps in circumstances similar to his dedication of the *Ikhtīyārāt*, as will be discussed in the next section.

---

[78]Dhahabī, *Tārīkh al-Islām wa-wafāyāt al-mashāhīr wa al-a'lām*, vol. 54, 101. It is conceivable that al-Dhahabī's words depict Shīrāzī at court with a youthful Öljeitü, prior to his accession in 1304 C.E.

[79]See Sheila S. Blair, "The Mongol Capital of Sulṭānīya, 'The Imperial'," Iran 24 (1986): 139–151. See also Minorsky, "Sulṭānīya." One way to interpret al-Dhahabī's remarks is to envision Shīrāzī as a tutor for the young prince; though this is, obviously, conjectural.

[80]Walbridge, "The Philosophy of Qutb al-Din Shirazi," 34–35.

[81]Walbridge also dates the authoring of Shīrāzī's work *Miftāḥ al-miftāḥ* to the period of the purported trip to Gīlān, though it is not clear if this is based on evidence from the *Miftāḥ al-miftāḥ*, itself. Walbridge, "The Philosophy of Qutb al-Din Shirazi," 33.

[82]Melville, "The Īlkhān Öljeitü's Conquest of Gīlān (1307): Rumour and Reality," 87.

3.9 The Patrons

Shīrāzī dedicates the *Nihāya* to "Muḥammad Ibn ... Bahā' al-Dīn Muḥammad Juwaynī."[83] Mishkat interprets this to mean Bahā' al-Dīn Muḥammad Juwaynī, the infamously harsh governor of Isfahān and *'Irāq-i 'ajam* and son of Shams al-Dīn the great Ṣāḥib Dīwān (which would make Bahā' al-Dīn nephew of the great historian 'Alā' al-Dīn Juwaynī).[84] Mir follows Mishkat in identifying Shams al-Dīn's son as the dedicatee of this work.[85] This identification is immediately problematic, however, due to the disagreement between the name of the dedicatee as it appears in Shīrāzī's book and the name of the candidate suggested by Mishkat and Mir. It is reasonably clear that Shīrāzī's patrons name was Muḥammad and that he was the *son* of Bahā' al-Dīn. There is, in addition, a chronological problem with the aforementioned identification, for the date of Bahā' al-Dīn's death is 1278 C.E./678 A.H., i.e., 3 years before the completion of the *Nihāya*.[86] The correct identification of the dedicatee of this work appears, instead, in Walbridge *The Science of Mystic Lights*. Noting the chronological problem with the dedicatee proposed by Mishkat and Mir, Walbridge proposes Shams al-Dīn the *Ṣāḥib Dīwān* himself as the dedicatee of this work.[87] As we saw in Chap. 2, Shams al-Dīn was put to death by Arghūn in Nov. 1284 C.E./Sha'bān 683 A.H., which would have been 3 years after the completion of the *Nihāya*. In addition Shams al-Dīn's given name was Muḥammad. Furthermore, his father is identified as Bahā' al-Dīn (the son of Muḥammad) by Spuler.[88] The definitive proof for Shams al-Dīn's identity as the dedicatee of the *Nihāya*, however, lies in Shīrāzī's dedication itself, for in it we also see included the name Shams al-Dīn and the title *Ṣāḥib Dīwān*.[89] Since according to al-Sallāmī, Shīrāzī worked as an assistant to al-Kātibī in Shams al-Dīn's school (presumably sometime in the period between 1269 and 1274 C.E.) he would likely have been a beneficiary of Shams al-Dīn's patronage for at least 7 years before the completion of his *Nihāya*, making the powerful administrator a natural choice as dedicatee.

Rather than a central and important figure such as Shams al-Dīn, the dedicatee of the *Ikhtīyārāt*, Muẓaffar al-Dīn Yavlaq (or possibly Yūlūq) Arslan, appears to

---

[83]Quṭb al-Dīn Shīrāzī, *Nihāyat al-idrāk fī dirāyat al-aflāk*, Köprülü MS 957, 1r.

[84]Shīrāzī, *Durrat al-tāj li-ghurrat al-dabāj*, n.

[85]Mir, *Sharḥ-i ḥāl wa āsār-i 'allāmah Qutb al-Dīn Maḥmūd Ibn Mas'ūd Shīrāzī, dānishmand-i 'ālī qadr-i qarn-i haftum, (634–710 A.H.)*, 70.

[86]Mudarris Razavi, *Aḥwāl wa Athār-i Muḥammad Ibn Muḥammad Ibn al-Ḥasan al-Ṭūsī*, 89.

[87]Walbridge, *The Science of Mystic Lights*, 181.

[88]Spuler, "DJuwaynī, Shams al-Dīn Muḥammad b. Muḥammad," in Encyclopaedia *of Islam, Second Edition*, Edited by: P. Bearman., 2010, http://www.brillonline.nl/subscriber/entry?entry=islam_SIM-2132.

[89]Shīrāzī, *Nihāyat al-idrāk fī dirāyat al-aflāk*, Köprülü MS 957, 1r. Indeed, the colophon of Köprülü MS 956 indicates that this work was completed in the very school that was founded by Shams al-Dīn in Sīvās. See note 1, Chap. 4.

have been a minor ruler from the somewhat peripheral Anatolian principality of Qasṭamūnī (i.e., modern Kastamonu which lies not far from the coast of the Black Sea, near Sinope). As we will see in the next chapter, it can be shown conclusively that the *Ikhtīyārāt* was completed shortly after the *Nihāya.* As we have already seen Shīrāzī had lived in Anatolia for some years prior to the completion of the *Nihāya.* It is also perhaps fair to assume that subsequent to the execution of Mu'īn al-Dīn in 1277 C.E., Shīrāzī's patron, Shams al-Dīn would have had an even more direct say in the administration of this vassal state of the Ilkhans, thus providing an opportunity (at least) for a strengthening Shīrāzī's ties with the local rulers. Still, the localities that are associated with Shīrāzī's stay in Anatolia (i.e., Sīvās, Malaṭīya, and even Konya) are at a fair geographical distance from Qasṭamūnī/Kastamonu, the seat of the dedicatee of the *Ikhtīyārāt.*

What reason, then, could compel Shīrāzī to dedicate his work to Muẓaffar al-Dīn? Cahen notes Kastamonu's "remoteness from the political centers" as an explanation for the lack of historical information regarding its establishment as a principality.[90] Indeed a comparison of the secondary literature indicates that there is disagreement even as to the name of the rulers of Kastamonu and their regnal years through the course of the thirteenth century.[91] The historical evidence such as it is, consists primarily of short entries in the history of Ibn Bībī and in Aqsārā'ī's chronicle (see Chap. 1, Sect. 1.3.2).

The most relevant account referencing the dedicatee of the *Ikhtīyārāt* appears at the end of the abridged version of Ibn Bībī's history, referred to in Chap. 1. Here we see Muẓaffar al-Dīn play a notable role in connection with the succession issues that faced the Seljuks in the aftermath of Baybars's campaign in Anatolia, c. 1275–1277 C.E. Up until Mu'īn al-Dīn's rule, the Mongols had successfully followed a shrewd policy of appointing rival Seljuk claimants to "rule" different parts of Anatolia.[92] Shortly after the coming to power, Mu'īn al-Dīn had managed to orchestrate a "unification" of the Seljuk territories, causing one of the pair of sultans ruling the Seljuk realms, 'Izz al-Dīn, to flee to Constantinople and then to the Crimean peninsula. The period from 1261 to 1277 C.E. had seen, therefore, the nominal rule of a single Seljuk ruler, the Sultan Rukn al-Dīn (presumably with Mu'īn al-Dīn Parvāne wielding actual power).[93] The events of 1275–1277 C.E. were concurrent with the violent uprising of numerous "Turcomen" entities.[94] These entities were generally Anatolian tribal groups, such as those ruling Kastamonu and other frontier

[90]Cahen, *Pre-Ottoman Turkey*, 310.

[91]Aḥmad Tauhid, "Rūm Seljuqu daulatinin inqiraz-ile teshkil eden tawa'if muluk (ma ba'd)," *Tarih 'Uthmani Encumeni Mecmu'esi* 5 (December 1910): 319; Eduard Karl Max von Zambaur, *Manuel de Généalogie et de Chronologie Pour l'histoire de l'Islam* (Bad Pyrmont: Orientbuchhandlung Heinz Lafaire, 1955), 148; Cahen, *Pre-Ottoman Turkey*, 310; O. Turan, "Anatolia in the Period of the Seljuks and the Beyliks," in *The Cambridge History of Islam*, vol. 1 (Cambridge: Cambridge University Press, 1970), 266.

[92]Cahen, *Pre-Ottoman Turkey*, 278.

[93]Cahen, *Pre-Ottoman Turkey*, 280. See also Chap. 2, Sect. 2.3.1.

[94]Cahen, *Pre-Ottoman Turkey*, 286–291; Melville, "Anatolia under the Mongols," 69–71.

areas that were often not under direct Seljuk rule. No doubt enticed by the mayhem (as well, perhaps, by tribal affiliations dating from the period leading to the ousting of his father) it was the son of the Seljuk Sultan ʻIzz al-Dīn, who in 1280 C.E. sailed across the Black Sea from Crimea and landed in Sinope with the goal of reclaiming his throne. Ibn Bībī recounts the allegiance of Muẓaffar al-Dīn, Shīrāzī's patron (and dedicatee of the *Ikhtīyārāt*), to the new claimant, Ghīyāth al-Dīn Masʻūd, as follows:

> The news reached Prince Muẓaffar al-Dīn Ibn al-Buyurk, whose ancestors had conquered and held those regions for generations, and he joined [the claimant] .... The Sultan [i.e., the claimant, and soon to be Sultan, Ghīyāth al-Dīn] ... added Prince Muẓaffar al-Dīn to his retinue and turned towards the great [Mongol] general Samāghār Bahādur who was the governor and the protector of the *limes* of Rūm. When he arrived, everyone – Mongol and Muslim – was struck by his comely face and all were impressed by his comportment and presence; and each [paid his respects] according to his abilities. The Mongol commanders dispatched Prince Muẓaffar al-Dīn as a member of his high retinue to the service of the threshold of the most high *ordū* [i.e., the Mongol court in Tabrīz], despite the fact that the host of winter was on the offensive and water ... had turned as stiff as a miser's hand, and in no time he was received at the glorious ... court. He was bequeathed prodigious and unanticipated honors and was granted the region of Amid [i.e., Diyarbakir] and the lands of Kharberd [i.e., modern Elâzığ] and Malaṭīya and Sīvās together with their citadels and their revenues, and was bolstered, as well, by many goodly promises.[95]

It is important to note here that the arrival of Muẓaffar al-Dīn at the Mongol court c. 1281 C.E. corresponds roughly with the completion of the *Nihāya* (i.e., Nov. 1281). If, as Aqsārā'ī states, Ghīyāth al-Dīn Masʻūd, was received by Abaqā, then his arrival at court (and that of his retinue, including Muẓaffar al-Dīn) would have occurred before Abaqā's death in April of 1282 C.E./20 Dhū al-Ḥijja 680 A.H.[96] In Chap. 4 we will see that the *Ikhtīyārāt* was completed less than 4 months after the *Nihāya*. Ibn Bībī's account of Muẓaffar al-Dīn's arrival at the Mongol court provides us, therefore, with an idea of how Muẓaffar al-Dīn came to be the dedicatee for the *Ikhtīyārāt*. If, we take the winter referenced in Ibn Bībī account to be that spanning December of 1281 to February of 1282, then Muẓaffar al-Dīn's arrival at court in Tabrīz would match the completion date of the *Ikhtīyārāt*. If, as Ibn Bībī suggests, Muẓaffar al-Dīn was granted Malaṭīya and Sīvās as fiefdoms, then this would not only have made the "Turcomen" amir from far-flung Kastamonu a good choice to serve as dedicatee for the *Ikhtīyārāt*, it would have also provided Shīrāzi and Muẓaffar al-Dīn ample opportunity to meet in Sīvās, itself.[97]

[95]Ibn Bībī, *Akhbār-i Salājiqah-i Rūm*, 337.

[96]Maḥmud ibn Muḥammad Aksarayi, *Tarikh-i Salājiqah, ya, Musāmarat al-akhbār wa musāyarat al-akhyār*, Majmuʻah-i tarikh-i Iran; 11; ([Tehran]: Intisharat-i Asatir, 1983), 134; Rashīd al-Dīn Ṭabīb, *Jāmiʻ al-tawārīkh*, 779.

[97]Though it is possible to proved an alternate reading to Ibn Bībī's account and assume that fragment is referring to Ghīyāth al-Dīn Masʻūd as the recipient of Sīvās, Malaṭīya, and Kharberd, this reading is considerably less probable. Ghīyāth al-Dīn Masʻūd was soon to be granted, as we will see, what was practically the entire Seljuk polity.

Rather remarkably, 679 A.H. (1280 C.E./1281 C.E.) is the year in which Mujīr al-Dīn Amīrshāh, the dedicatee for Shīrāzī's third book on astronomy (*the Tuḥfa*) rose to prominence at Mongol court, as well. Assuming the post that had belonged to his father, Tāj al-Dīn al-Mu'tazz, i.e., the Mongol-appointed financial officer who oversaw the repayment of Seljuk loans dating from Hülegü's campaign some 20 years prior[98]

> Mujīr al-Dīn Muḥammad Ibn al-Mu'tazz came to Rūm and by virtue of the *yarlighs* and the *paizas* obtained in the year 679 A.H. from ... Abaqā with royal honors, revived the position of his father, and took control of the *īnjū* and *muqāṭa'āt* of the kingdom that had been earmarked for the treasury of the High Presence [of the royal court] as well as the *bālish*. And, verily, the kingdom flourished greatly through his constructive efforts.[99]

A description of the various forms of state revenue listed in the fragment appears in Cahen.[100] *Muqāṭa'āt* refers to regions leased as tax farms or for natural resources, and *īnjū* to lands that belonged to the state (and that provided revenues to the central treasury). The meaning of the term *bālish* is not known precisely, though it is obviously a form of tax or tribute.

It should be noted here that Mujīr al-Dīn's position obtained via Abaqā's decree granted him control over the principality of Kastamonu among other locales, and thus Muẓaffar al-Dīn's appearance at court (as a partisan of Mas'ūd's claims to the Seljuk throne) could hardly have been a mere coincidence.[101] Was the presence of the amir from the autonomous region of Kastamonu in the entourage of the claimant to the Seljuk throne cause for concern at the court at Tabrīz? Was Muẓaffar al-Dīn's divestment of his ancestral territory and his relocation to Sīvās – considerably closer to Tabrīz than Kastamonu – carried out, perhaps, in order to keep a closer eye on him? The answer to these questions can not be provided at present, but the dynastic claims of Ghīyāth al-Dīn Mas'ūd, the arrival of Muẓaffar al-Dīn at the Mongol court, and the rising fortunes of Mujīr al-Dīn occur close enough in time, as to suggest that they were in some way closely and causally linked.

By the time of the dedication of the *Tuḥfa* Mujīr al-Dīn had seen a steady increase in his fortunes. Abaqā's successor, Tegüder Aḥmad, had decided to revert to the time-tested Mongol strategy a divided Anatolia, appointing Ghīyāth al-Dīn Mas'ūd as ruler to the traditional realms of the Seljuk polity, and re-assigned the existing Sultan, Ghīyāth al-Dīn Kay-Khusrau, as ruler of the southern coast of Anatolia. Mujīr al-Dīn had been assigned as Ghīyāth al-Dīn Mas'ūd's deputy, or the *nā'ib al-salṭana*.[102] Given the fact that this appointment would have made him one of the most powerful men in Rūm (with an influence perhaps rivaling that of Shīrāzī's deceased patron, Mu'īn al-Dīn Parvāne) his choice as dedicatee of a major scientific work, is, therefore, not surprising.

---

98 Cahen, *Pre-Ottoman Turkey*, 332.

99 Aksarayi, *Tarikh-i Salājiqah, yā, Musāmarāt al-akhbār wa musāyarāt al-akhyār*, 134.

100 Cahen, *Pre-Ottoman Turkey*, 333.

101 Cahen, *Pre-Ottoman Turkey*, 332.

102 Melville, "Anatolia under the Mongols," 73.

3.10 Shīrāzī's Patrons: Concluding Remarks

Shīrāzī's choice for the dedicatee of the *Ikhtīyārāt* allows us to draw a provisional conclusion in regard to his choice of patrons in general. It appears as if, of Shīrāzī's three works on astronomy, the ones that were written in Arabic were in turn dedicated to administrators of the Ilkhan court who, as bureaucrats would have had a firm grasp of Arabic themselves. The *Ikhtīyārāt* is dedicated to a ruler of Kastamonu who in all likelihood had a limited ability in Arabic. Indeed, that Muẓaffar al-Dīn was able to read the *Ikhtīyārāt* in Persian (which would have been quite different from his native Turkish) indicates that he was a fairly educated man. Here, it is also difficult not to recognize a parallel between the dedication of the *Ikhtīyārāt* and that of the *Durra*. Like the *Ikhtīyārāt*, the *Durra* is dedicated to a minor ruler (rather than a powerful administrator working for the Ilkhanid state). There can be little doubt that the abilities of "Amīra Dabāj" from the relative backwater of Gīlān would have been as limited as that of Muẓaffar al-Dīn with regard to facility in Arabic.

Little else is known of the details of the relationship between scholars such as Shīrāzī and the patrons/dedicatees of his scholarly works. For a high official to serve as the dedicatee of a book would no doubt have involved gifts, career appointments, and stipends for the author. At least as important would have been the social capital that powerful patrons such as Juwaynī could have provided their clients such as Ṭūsī and Shīrāzī. This capital would have taken the form both of the added prestige and authority granted to the client by virtue of his association with a famous patron (who in turn would have had a long list of outstanding scholars as clients) as well as by providing a space in which to continue ones scholarly pursuits relatively unhindered (i.e., by providing protection from competing scientists and enemies at court). The patron in turn would have benefited from the prestige associated with the support of knowledge, culture and learning.

In regard to those of Shīrāzī's patrons we have seen so far, one may assume that the works that he rendered in Persian (the *Ikhtīyārāt*, and the *Durra*), provided an opportunity for patrons from peripheral areas to obtain the prestige of patronizing men of letters such as Shīrāzī. Unlike the longstanding support of powerful statesmen such as Juwaynī (with its prestige and career-enhancing qualities) it is difficult to see, however, how benefits other than monetary stipends and gifts, could have accrued to the author.

As a brief review of this chapter, it should be noted that a comparative study of the works of Ibn al-Fuwaṭī, al-Dhahabī, al-Sallāmī, and al-'Asqalānī together with Shīrāzī's autobiographical notes allows us to trace a rough trajectory of Shīrāzī's whereabouts through his life as a scholar. This trajectory would have taken Shīrāzī from his birthplace to Marāgha in 1259–1260 C.E, and then to Juwayn, Qazwīn, and Baghdād in the subsequent period of a little over a decade. Shīrāzī appears to have then moved to Anatolia, for he tells us of his residence in Konya in 1274 C.E. The historical accounts reviewed in this chapter suggest strongly that his appointment as judge in Sīvās would have occurred shortly after 1274 C.E. for one of his main benefactors, Mu'īn al-Dīn (whose name is associated historically with Shīrāzī's appointment) was executed in 1277 C.E. Shīrāzī would presumably have remained in Sīvās until the completion of the *Ikhtīyārāt*. As we will see in the next chapter this would have occurred shortly after the completion of the *Nihāya* (i.e., Nov. 1281 C.E.).

The ensuing period appears to have been especially hectic. Shīrāzī left Sīvās for Tabrīz, perhaps upon the death of Abaqa, in April 21st 1282 C.E., and shortly thereafter went on his embassy to Cairo (Aug. 1282 to Jan. 1283). Yet, by August 1285 C.E., Shīrāzī was back in Sīvās where he completed the *Tuḥfa*. In all likelihood the events surrounding the death of Tegüder Aḥmad (August 1284 C.E.) and especially the execution of Shīrāzī's benefactor Shams al-Dīn Juwaynī (October 1284 C.E.) were at least partly responsible for this move from Tabrīz to Sīvās. In the absence of other historical data we may reasonably wonder if the account of Shīrāzī's meeting with Arghūn in Anatolia c. May 1290 C.E., does not in some form signify his coming back into favor at the court in Tabrīz. Whatever, the case may have been the sources that we have looked at are unanimous in stating that Shīrāzī spent the last decades of his life in Tabrīz, and that he was buried there. Based on the information from Shīrāzī's autobiography, the period following Ghāzān's accession in 1295 C.E. was particularly stable and pleasant for him. The historical narratives we have studied mention that he was busy with his scholarship, both as an author and as a teacher, during this period. Several of his books, including the *Durra*, are the fruits of this late period in Shīrāzī's life.

Chapter 4
The Principal Astronomical Sources

4.1 Chronological Considerations

Existing manuscript copies that have reached us of both the *Nihāya* and the *Tuḥfa* include the dates in which these works were completed. The colophon from Köprülü 956 indicates that the *Nihāya* was completed in the middle of Sha'bān in 680 A.H., corresponding to late November or early December in 1281, C.E. in the city of Sīvās.[1] This date is repeated in Köprülü 957, which references a work written in the author's own hand.[2] BN Arabe 2516, a manuscript of the *Tuḥfa* indicates that this work was also completed in Sīvās and that the date of its completion was in August, 1285 C.E.[3]

In contrast, none of the copies of the *Ikhtīyārāt* originally examined for this study listed a completion date for the work, so the precise dating of it presented a problem. One of the few earlier attempts to date this work was made by Saliba in "Persian scientists in the Islamic world." In this essay Saliba assigns a rather late date to this work, due to his assumption that the *Ikhtīyārāt* must have been written after the *Tuḥfa*.[4] However, evidence from the manuscripts themselves refutes this, and suggests that the *Nihāya* and the *Ikhtīyārāt* were written in close succession.

One of the most convincing bits of evidence for the temporal proximity of these two works is the reference by name to the *Ikhtīyārāt* in MS Köprülü 957 and MS Köprülü 956 – early manuscript copies of the *Nihāya*, that were completed

---

[1] Shīrāzī, *Nihāyat al-idrāk fī dirāyat al-aflāk*, Köprülü MS 956, 148r. This colophon indicates that the manuscript was completed in the madrasa founded by Juwaynī in Sīvās. This madrasa is in all likelihood the Çifte Minare Medresesi (see Sect. 3.6).

[2] Shīrāzī, *Nihāyat al-idrāk fī dirāyat al-aflāk*, Köprülü MS 957, 195r.

[3] Shīrāzī, *al-Tuḥfa al-shāhīya*, BN Arabe MS 2516, 118r.

[4] Saliba, G. "Persian scientists in the Islamic world." In *The Persian Presence in the Islamic World*, Richard G. Havannisian and Georges Sabagh, Ed., Cambridge: Cambridge University Press, 1991, 138.

K. Niazi, *Quṭb al-Dīn Shīrāzī and the Configuration of the Heavens: A Comparison of Texts and Models*, Archimedes 35, DOI 10.1007/978-94-007-6999-1_4,

on August 30th, 1282 C.E., and on December 20th, 1284 C.E.[5] The name of the *Ikhtīyārāt* appears clearly in the margin of folio 72r of MS Köprülü 957, suggesting that the note was added later, i.e., that it was originally missing from the *Nihāya* (which had been completed about 9 months prior). In MS Köprülü 956 the reference to the *Ikhtīyārāt* again appears plainly, on folio 58v. What was a marginal note in MS Köprülü 957 has moved here to the body of the text itself. Given the possibility that the marginal notes in MS Köprülü 957 were added after the work was completed, the evidence indicates December 20th, 1284 C.E. (i.e., the completion date for MS Köprülü 956) as an upper limit for the completion of the *Ikhtīyārāt*. In the absence of additional evidence, therefore, one could safely date the *Ikhtīyārāt* to the period between November 1281 (i.e. the completion date for the *Nihāya*) and December 1284.

Since the original publication of these results additional evidence has been fortuitously forthcoming, however. Recently a dated manuscript copy of the *Ikhtīyārāt*, MS Milli Library 31402, was brought to light by Gamini. This manuscript allows us to date the *Ikhtīyārāt* precisely, and confirms the proximity of its completion date with that of the *Nihāya*. Based on MS Milli Library 31402, the *Ikhtīyārāt* was completed on the 9th of Dhū al-Ḥijja, 680 A.H. (Feb. 19th, 1282 C.E.) in Sīvās.[6] We can see therefore, that Shīrāzī finished the *Ikhtīyārāt* less than 4 months after the *Nihāya*.

A brief review of the discussion so far in regard to the chronology of the three astronomical works in question indicates the following: After completing the *Nihāya* (in November, 1281 C.E.), Shīrāzī appears to have started on the *Ikhtīyārāt*, completing it in February, 1282 C.E. This was coincident with the arrival, at the Ilkhan court in Tabrīz, of Muẓaffar al-Dīn, the dedicatee of this work (and with his being granted the rule of Sīvās). During the copying of MS *Nihāya* Köprülü 957 a little while later, Shīrāzī appears to have inserted the reference to the *Ikhtīyārāt* in the margins; a reference that in later manuscript copies of the *Nihāya* was incorporated into the main text. Several years later, in August 1285 C.E., Shīrāzī completed yet another work on *hay'a*, i.e., the *Tuḥfa*.

As far as his authoring of astronomical works is concerned, Shīrāzī's apparent productivity in the period spanning 1281–1285 C.E. is rather striking. What were Shīrāzī's motives and reasons for writing three major works on the same topic in such rapid succession? Was Shīrāzī trying to say something new in each successive work, or was he merely restating the same information? If these works are not mere repetitions of each other, then in what ways are they different? In order to answer these questions and the question of "Why write three books on Astronomy topic in the span of less than four years?" we now turn to look more carefully at the books themselves.

---

[5]Shīrāzī, *Nihāyat al-idrāk fī dirāyat al-aflāk*, Köprülü MS 957, 195r., Shīrāzī, *Nihāyat al-idrāk fī dirāyat al-aflāk*, Köprülü MS 956, 148v.

[6]I gratefully acknowledge Dr. Gamini's generous help, in communicating his crucial find, and providing me with digitized images of the relevant folios of MS Milli Library 31402.

4.2 The Exordia

In this section we will look at the introductory (and the closing) sections in the *Nihāya*, the *Tuḥfa*, and the *Ikhtīyārāt* in order to better understand the genesis of these books and the relationship between them. Though each of the examined fragments is replete with conventional tropes and flourishes, it is possible to glean useful information while trying to read between Shīrāzī's conceits and his fanciful figures of speech. In the introduction to the *Nihāya* Shīrāzī states:

> I had wished for a period of time to compose for myself and my brethren in the science of *hay'a* ... a self-sufficient epistle, inclusive of the cream of the written explications and the pith of the collected discourses on the composition [tarkīb] of the orbs, containing a summary of what has been achieved and the results of that which the utmost of attainment has reached, in order for this work to be a demonstration for the beginner as well as a reference for the expert; [and even more so] the foundation for those of perspicuity and the utmost limit of this science for those of intelligence; yet lesser obstacles had prevented me....[7]

According to Shīrāzī's statement, then, the *Nihāya* was meant as a primer for the beginning astronomer as well as a work of reference for the more accomplished practitioners of astronomy.[8]

This introductory fragment contains a rhetorical flourish that would have been apparent to knowledgeable readers, for, embedded within it are references to astronomical works that are the predecessors of the *Nihāya*, and from which Shīrāzī's work presumably draws. This reference subtly reinforces Shīrāzī's claims as to the comprehensive nature of his work, as well as positing its excellence relative to the other well-regarded works of his era.[9] A bit further in the introduction Shīrāzī states:

> When I asked the Lord for guidance and commenced in the composition of this book, a person from whom I am unable to withhold a favor and one whom I am unable to contradict, being the dearest of my friends and the foremost among them in virtue, Muḥammad ibn ʿUmar al-Badakhshānī ... requested from me that, where necessary, a gentle indication be made of [the method of] observation and an amiable sign made of the manner of

---

[7]Shīrāzī, *Nihāyat al-idrāk fī dirāyat al-aflāk*, Köprülü MS 957, 1v.

[8]See the discussion in Ṭūsī, *Naṣīr al-Dīn al-Ṭūsī's Memoir on Astronomy*, 37, in regard to the *Tadhkira* having been written, in part, with the student of astronomy in mind. It is also worth noting here the striking contrast between Ṭūsī's laconic style in his introduction relative to Shīrāzī's verbosity.

[9]There are references to the following works, emphasized as well at the conclusion of the *Nihāya*. Not all of these works can be identified. The list of references consists of *al-Mughnīya* (perhaps the *Muʿīnīya*, by Ṭūsī), *al-Zubda* (perhaps *Zubdah-i hay'a* or *Zubdat al-idrāk fī al-hay'a* by the same author), *al-Lubāb* (?), *Ghāyat al-afkār* (?), *al-ʿUmda al-ūlaa* (?), *al-Mulakhkhaṣ* (?), *Tarkīb al-aflāk* (perhaps *Kayfīyyat tarkīb al-aflāk*, by Jauzjānī; the author of this book is mentioned unkindly in several of Shīrāzī's works), *al-Tadhkira* (by Ṭūsī), *al-Muḥaṣṣal* (?), *Muntahā al-idrāk* (perhaps *Muntahā al-idrāk fī taqsīm al-aflāk*, by Kharaqī, *al-Tabṣira* (perhaps *Kitāb al- Tabṣira fī ʿilm al-hay'a*, also by Kharaqī, another one of the authors mentioned by Shīrāzī in his astronomical works). Shīrāzī, *Nihāyat al-idrāk fī dirāyat al-aflāk*, Köprülü MṢ 957, 1v.

> extracting the motions ... and that I study the words of the *Tadhkira* which has no equal from among its precedents and which will remain unsurpassed by its successors, and to insert these in my words should the import be apparent and to simplify them if there is a species of obscurity in them. And I have met his prescription and realized his hope [thus collecting both advantages].... And since this book has not "left out anything great or small, but takes account thereof" [*The Qur'ān* 18:49] and since there does not exist a haughty pronouncement or an uncouth one without [my book] ridiculing or belittling it, [doing so] by encompassing the foremost thoughts of the ancients and comprehending the limits of the views of the others from among the moderns together with noble benefactions and refined pearls originating from us – and if it is not more glorious than what we have mentioned and greater, it is not lesser – I have called it the "Limits of Attainment in the Understanding of the Orbs," in order that its name be a guide to its import and in order that its appearance bear news of its meaning and I have arranged it in four sections ... and to God I pray humbly for the completion of that which I have set as my goal.[10]

Shīrāzī thus describes the genesis of his *Nihāya* as a commentary on the *Tadhkira* of his teacher Ṭūsī (which according to Ragep became one of the most important works in *hay'a* subsequent to its publication).[11] Indeed, as we will see, portions of the *Nihāya* – the chapter on the superior planets, for example – consist of Ṭūsī's words interspersed with explanations and clarifications by Shīrāzī, thus complying with the basic format of a commentary. Shīrāzī does not consider himself as bound by the material in the *Tadhkira*, however. Indeed, entire sections of the *Tadhkira* are not referenced at all in the *Nihāya*.[12] Shīrāzī appears for much of the book to have followed Ṭūsī's conception of producing a primer for *hay'a*, but was also yearning to improve and to supplant scientific models of his predecessors, Ṭūsī, and al-'Urḍī.

In the conclusion of the *Nihāya* Shīrāzī writes:

> And this is the end of the book and thanks be to God, the inspirer of judgement, for this is what was allowed by my disposition and thoughts, wounded as they were by the knocking about of the years ... and uncountable preoccupations ... while I was exerting the limits of my power in the uncovering and the rendering of meanings together with the abridgment and summarizing of their rules. And I produced solutions that had not occurred to anyone prior to me....[13]

The solutions that Shīrāzī alludes to are a reference to his proposed configurations for the orbs of the planets. In his claim to having produced novel results Shīrāzī is conjuring the perceived inadequacies with Ptolemy's work that were the driving force for the science of *hay'a* in the Islamic world. In addition to affirming Shīrāzī's notion of having succeeded where others had failed, the subsequent text asks for a fair assessment of his work from his readers:

---

[10] Shīrāzī, *Nihāyat al-idrāk fī dirāyat al-aflāk*, Köprülü MS 957, 2r.

[11] Ṭūsī, *Naṣīr al-Dīn al-Ṭūsī's Memoir on Astronomy*, 55.

[12] Ṭūsī's discussion of his implementation of the "Ṭūsī couple" in the configuration of the planetary orbs, which occupies a good portion of Book II, Chapter 11 of the *Tadhkira* is only referenced, for example, in the briefest fashion, allowing Shīrāzī to champion alternative models instead. For a discussion of the importance of commentaries as a genre see Ṭūsī, *Naṣīr al-Dīn al-Ṭūsī's Memoir on Astronomy*, 59, and Saliba, *Islamic Science and the Making of the European Renaissance*, 241.

[13] Shīrāzī, *Nihāyat al-idrāk fī dirāyat al-aflāk*, Köprülü MS 957, 197r.

> And I beseech the reader of my book to avoid hastening toward the rejection of that with which he is not familiar or that which is opposed to his nature; rather it is incumbent upon him to look intently [at the book] and to avoid being inconsiderate, and subsequently to follow the path of denial or of admission ... and [also I ask] that he correct what has befallen it [i.e., the book] as far as faults and corruption ... And that he remember me with his most honest prayer .... As Aristotle says in his *Metaphysics*, it is not meet to thank him who says much in regard to the Truth; rather it is fitting to thank him who says little. This despite the fact that that which we have said is not inferior to what [our predecessors] have said nor is it lesser; it is superior, rather, and greater.[14]

Original theoretical models in *hay'a*, concerned as it was with the structure of the cosmos itself, could not have avoided the creation of a certain amount of tension due to their differences with the authoritative models of Ptolemy.[15] The principles of natural philosophy as understood by Shīrāzī and his colleagues, however, appear to have provided a sanctioned approach to the production of original models that departed from those Ptolemy had produced. This is suggested in the same section of the *Nihāya* itself; for, following his appeal to his colleagues for an unbiased appraisal, and subsequent to re-iterating the list of authoritative books alluded to in the introduction, Shīrāzī invokes the principles of *hay'a*:

> And perhaps this can be understood by reading the well-regarded books composed on this topic, some of which have been indicated in the introduction to this book; and by comprehending their meanings and understanding the fundamentals of their principles and then by a comparison between them and this book in order to distinguish the chaff from the kernel. And the Lord is the inspirer of truth and well-guidedness. From Him is the beginning and to Him is our return. And since God has granted me the completion of what I had intended ... we end the book thanking God....[16]

Shīrāzī's statement, that his cosmological models and those of his colleagues should be judged by "the fundamentals of their principles" is important in that it refers to one of the main preoccupation of the *hay'a* authors, i.e., the desire to render the workings of the universe in a manner consistent with a guiding set of principles grounded in natural philosophy.

Though Shīrāzī's other major work in Arabic under consideration here, the *Tuḥfa*, was written on the same topic as the *Nihāya*, Shīrāzī's ostensible aims, as expressed in his two introductions appear to be somewhat different. As we saw, the *Tuḥfa* was completed in Sīvās after a period of a little less than 4 years after the completion of the *Nihāya*. We know from historical sources that Shīrāzī was preoccupied with

---

[14]Shīrāzī, *Nihāyat al-idrāk fī dirāyat al-aflāk*, Köprülü MS 957, 197r. The quote from the *Metaphysics* remains unidentified.

[15]By exhorting the reader to not judge his models too hastily, Shīrāzī may also have been acknowledging the difficulties in proposing planetary configuration different from what appeared in the authoritative tradition of Ptolemy. However, by invoking Aristotle's authority immediately prior to his confident claims in regard to his own innovative work in astronomy Shīrāzī appears to be hearkening to an even greater authority on physical theory, i.e., Aristotle, from whom the principles of *hay'a* and of natural philosophy ultimately derived. See Saliba, "Aristotelian Cosmology and Arabic Astronomy," 251–268.

[16]Shīrāzī, *Nihāyat al-idrāk fī dirāyat al-aflāk*, Köprülü MS 957, 197r.

affairs other than the writing of astronomy books for much of this period. Among other things he traveled to Cairo on a diplomatic mission and (while in Cairo) eagerly searched for books to aid him in the completion of his commentary on the first book of Avicenna's *Canon*.[17] In addition two of his patrons – Shams al-Dīn Juwaynī, to whom, as we saw, the *Nihāya* was dedicated, and the Ilkhan ruler Tegüder Aḥmad – were executed during this period. In the introduction to the *Tuḥfa* Shīrāzī states:

> Verily the neediest of God's creatures Maḥmūd ibn Mas'ūd al-Shīrāzī, may the Lord make his fate a blessed one, says if it weren't for the convention that permits the lesser to [offer supplication] to the greater then it would be the sanctuary of their company, the dependence upon their strength, and the pride in associating with them, [and independence through reliance upon them] that would compel the weak to seek this association. When I discovered this custom ... and commenced on observing this tradition, seeing fit according to the bestowal of gifts to the kings of one of the two states [mulūk iḥdā al-daulatayn] to bestow upon ... the son of Mu'taz ibn Ṭāhir ... Mujīr al-Dīn Amīr Shāh ... a gift that would remain for eternity and that would not be diminished by the passing of the years and months.[18]

One is struck here by the difference in tone relative to Shīrāzī's introduction in the *Nihāya.* While in his first *hay'a* book Shīrāzī highlights his selfless intentions in publishing the result of his astronomical studies, what is given prominence in the *Tuḥfa* is the relationship of the author with his patron Mujīr al-Dīn. The execution of Shams al-Dīn Juwaynī, less than a year before the completion of the *Tuḥfa* must have been a preoccupation of the author as he was writing this work, and this may be what gives his depiction of the protective attributes of the patron an added urgency in the introduction to this work.

In order to provide a motive for his *choice* of offering to his powerful patron, Shīrāzī proceeds by extolling his patron Mujīr al-Dīn's love of knowledge. "And since I had seen that knowledge was to him what was most desirable and the most glorious of gifts before him, I chose from it the science of *hay'a*, which praises the revelation sent to His two worlds, by virtue of His glorious words: 'Those who mention God standing and sitting and recumbent upon their sides, thinking about the creation of the heavens and the earth. The Lord has not created these in vain (*The Qur'ān* 3:191).'"[19] In a passage that parallels a similar one in the *Nihāya* (but is considerably shorter) Shīrāzī also praises the virtues of astronomy as the most excellent science and then resorts to the same rhetorical device as in the *Nihāya*, embedding the precedent for the *Tuḥfa* in the language describing the nature and content of the book itself:

> And I composed a book to appear with his name and with excellent words and principles and the most elegant discourses including the indications to treasures that are the glimmers of the "limits of attainment" and hints to the secrets that are the glances of the "understanding

[17]Walbridge, "The Philosophy of Qutb al-Din Shirazi," 23.

[18]Shīrāzī, *al-Tuḥfa al-shāhīya*, BN Arabe MS 2516, 1v.

[19]Shīrāzī, *al-Tuḥfa al-shāhīya*, BN Arabe MS 2516, 1v.

> of the heavens" while both assailing the falsehood in this art, and turning away from that which is subordinate to Truth; [our book rather] being confined to that which has resulted from our thoughts, and that upon which our opinion has settled, with no calumniation against the books of our companions, for there is nothing better than the urging of opposition in error toward agreement in the correct method.[20]

That, according to Shīrāzī's claim, criticism of the faulty work of other astronomers is not included in the *Tuḥfa* hints at a notable difference between this book and its predecessor, the *Nihāya*. This point is more clearly stated in the subsequent text:

> Let it be known that if something has not been mentioned in it, it will have been mentioned in the books of our companions, furthermore its omission here is due to its being disparaged by people of understanding; and it is incumbent upon him who wishes to be informed of it to refer to our book entitled "*The Limits of Attainment in the Knowledge of the Heavens*" in order to distinguish with it the kernel from the chaff and lo, I start the book called the *Royal Offering* organized in four chapters, asking the Lord, the inspirer of what is good, to aid in its completion, for verily He is the granter of success and to Him is our return.[21]

In addition to highlighting the *Nihāya* as a work of reference, Shīrāzī thus repeats his claim in regard to the *Tuḥfa* and how this work was meant to present only the state of the art, and that refuted theories, some of which may have been included in the *Nihāya*, have been omitted.[22] We will have an opportunity to return to the discussion of the *Tuḥfa* and its contents in a subsequent section of this study.

In the conclusion of the *Tuḥfa*, Shīrāzī alludes to the practical and utilitarian qualities of this work and refers yet again to the vicissitudes of fate, listing his duties as judge and teacher as having served as distractions during the composition of the *Tuḥfa*:

> And this is what was allowed by my dull disposition and my abated understanding in the midst of what I was obliged to face as far as the circumstances of loathsome affairs and the perseverance of irksome worldly pursuits, including jurisprudence and teaching.... I have offered it as a token of service to his highness of the treasury of the great lord and the noblest master and as a gift to his [noble] presence. May the Lord preserve his [protective] shadow upon the entirety of his servants and his clients .... And I hope that this servant's book falls into favor and that he, glory to him, is capable of obtaining his desire in regard to his fine pursuits ... and to that which he yearns for. And I hope to God that he grants success to the [reader] so that the benefits of [the book] can come to him, and that [the reader] forgives me for an oversight should he encounter it, for I myself am dismayed by my errors and recognize my feebleness. And thanks be to God who guided us to this, for we are unguided unless God guides us....[23]

Shīrāzī concludes the *Tuḥfa* with the time and place at which he completed this work.

---

[20] Shīrāzī, *al-Tuḥfa al-shāhīya*, BN Arabe MS 2516, 1v.

[21] Shīrāzī, *al-Tuḥfa al-shāhīya*, BN Arabe MS 2516, 1v.

[22] That these would have been included in the *Nihāya* as a more comprehensive reference work for the astronomer is understandable.

[23] Shīrāzī, *al-Tuḥfa al-shāhīya*, BN Arabe MS 2516, 119r.

It should be noted here that, as Shīrāzī states, the very name of his book is a reference to the name of the dedicatee, Mujīr al-Dīn Amīrshāh, by the inclusion of the word *shāh* (king, in Persian). This only serves to highlight the conception of the book, if not as a commissioned work, then at least as one that is closely linked with the client-patron relationship between the author and the dedicatee. The *Tuḥfa* shares this feature with Shīrāzī's other work under consideration in this study, the *Ikhtīyārāt*, since the name of the dedicatee of the work (Muẓaffar al-Dīn) appears in the title for the work itself, i.e., *Ikhtīyārāt-i Muẓaffarī*).[24]

Shīrāzī begins his Persian work on *hay'a*, the *Ikhtīyārāt*, with a somewhat ornate invocation:

> Untold thanks and adoration is meet to the ... Builder who has adorned the glass vessel [of the sky] with the gleaming pearls of the stars and the blazing jewels of the planets ... the Sage who has placed the scabbard of the sword of vengeance in the clasp of Saturn, the Savant who has sheathed Jupiter with a cloak of prosperity in the seat of lordship, the Victor who has appointed Mars as sheriff in the fifth realm, the Sovereign who through the gilt disk of the Sun – which is as the pupil of the entirety of creation –has illuminated the upper and lower parts of the metaphorical world, the Beneficent who has placed the organon of arts beside famed Venus, so that the itinerant Moon has fashioned its melodies into his dervish's cloak, the Ruler who has placed the pen of administration in the hands of Mercury, who is the composer of the second realm,... the King whose wizard-like might has tossed seven pairs [!] of gilt dice in this azure bowl, and has set thousands of crystal game-pieces in the twelve mansions of this kohl-darkened plot, so that through their influence his geometer-like wisdom could, at times, set the token of the actions and of the appointed times for the creatures of the world moving gainfully in creation and existence and, at other times, to have these be still in the realm of death and nonexistence. For creation and dominion are His alone. May the Lord, this most excellent of creators, be blessed.[25]

After this florid passage, Shīrāzī proceeds by praising astronomy and criticizing Ptolemy while alluding to the considerable effort expended by Shīrāzī's predecessors in ridding the Ptolemaic system of its perceived flaws.

> So says the author of these lines ... [Shīrāzī] that since the noblest kind of mathematics – which is a part of the theoretical sciences – is the science through the acquisition of which the human soul is ennobled by the knowledge of the configuration of the heavens and the Earth and the number of the orbs and the magnitude of the motions and the extent of the distances and the bodies and the situation of the simple bodies that are parts of this world, a fair portion of my life was spent in its pursuit and explication. And since that science, in the manner in which the expert in this art, the master of the *Almagest* has described was not devoid of great difficulties and the pre-eminent ones and the moderns ... had assiduously exerted a great deal of effort in resolving the problems and uncovering the intricacies, and had come up short – resorting to various tricks and innovative rules, some reversing the directions of the motions from that which the master of the *Almagest* had

[24]Both of these works, then, stand in contrast to the *Nihāya*, the title of which does not allude to the patron's name.

[25]Shīrāzī, *Ikhtīyārāt-i Muẓaffarī*, Ayasofya MS 2575, 1v. It is worth noting here that this lyrical invocation of the heavens and their effects on the creatures of the world, does not bear a resemblance to the manner in which Shīrāzī reflects on the relationship of clientage with his courtly patron in the opening of the *Tuḥfa*; again suggesting Shīrāzī's preoccupation in the cultivation of new relationships of patronage as part of his project for the *Tuḥfa*.

> stated and some leaving them as they were, yet all of them increasing the confusion of the orbs. And truth be told, to a person, none could fulfill this duty or emerge from within its strictures, some by their own admissions and some according to our inference as to the corruption of their physical laws, and since the [arm of victory] was adorned by the blessing of divine endorsement and the visage of that which was yearned for was embellished by the necklaces of Godly benefaction and the cloak of anticipation and mask of concealment removed from the countenance of the aims of the author of these lines, so that the solution of those problems were facilitated for him—whether through consulting the books of the experts of the art or through induction and the application of thought and vision, he desired, for the purpose of the safeguarding of excellence and the participation of other seekers ... to publish it and to preserve it from obliteration and dispersal, and to present it to the seekers of the [true] path and betterment who have set their wills to the search for truth.[26]

Shīrāzī's direct allusion to Ptolemy does not occur in either of the two other *hay'a* works under consideration. Furthermore, while the increase "in the confusion of the orbs" parallels Shīrāzī's claims of success in theoretical astronomy in the other two works, here his description of the failings of his predecessors is more detailed, referring clearly to the activity directed toward describing and predicting planetary motions.[27] Finally this passage is both a concession to one of the primary wellsprings of the *hay'a* tradition (i.e., the *Almagest*), as well as an explicit description of one of the main driving forces behind the works of Shīrāzī and his fellow astronomers in the Islamic world, namely that they viewed Ptolemy's work as faulty and in need of emendation.[28]

Shīrāzī then alludes to his success in treating the theoretical problems that have stymied his predecessors while (at the same time) ascribing the genesis of *Ikhtīyārāt* to the request of his patron:

> and by virtue of this inducement [i.e. his desire for safeguarding his findings] he composed the book "the Limits of Attainment in the Understanding of the Orbs" and due to the fact that that book included the limits of the thoughts of the ancients and the farthest extent of the views of the moderns and [since] for the purposes of the beginner the criticism and dispraisal of each of these and the recognition of that which is the preferred method from that which isn't appeared difficult, this was the inception of a mental disquiet regarding the need for preserving the preferred method and the summary of its secrets. During these thoughts ... there transpired an indication by ... [Muẓaffar al-Dīn] towards this sincere supporter and blameless adherent to arrange some chapters on the description of the orbs and the bodies and to beautify the ink of the explication of those inviolate meanings with Persian expressions so that they may benefit those of high rank and low ....[29]

According to Shīrāzī's introduction, therefore, the *Ikhtīyārāt* was written subsequent to the *Nihāya* and that the work served a double purpose: both to preserve for

[26]Shīrāzī, *Ikhtīyārāt-i Muẓaffarī*, Ayasofya MS 2575, 2r.

[27]By referring to the reversal of the motions of the orbs relative to the *Almagest*, Shīrāzī is likely referring to al-'Urḍī and his model for the Moon. See Saliba, "Arabic Planetary Theories after the 11th Century AD," 93.

[28]Abū ʿAlī al-Ḥasan Ibn al-Haytham, *Shukūk 'alá Baṭlamyūs* (al-Qāhirah: Maṭba'at Dār al-Kutub, 1971), 5; George Saliba, *Islamic Science and the Making of the European Renaissance*, 94–117.

[29]Shīrāzī, *Ikhtīyārāt-i Muẓaffarī*, Ayasofya MS 2575, 2r.

the beginner what Shīrāzī considers the "preferred" method and to preserve this knowledge in Persian (a language with which the patron of this particular work would likely have had a greater facility than with Arabic). Shīrāzī proceeds, as he did in both of his other works offering the same conceits of humility and meekness, begging forgiveness for the inadequacies of his book before starting his discussion of astronomy proper.[30]

In the conclusion of the *Ikhtīyārāt* Shīrāzī again formally asks the patron to overlook the faults of the book, and echoes his wish in the *Tuḥfa* regarding the usefulness of the book for the practical aims of the patron:

> And as what we promised in the introduction of the book has been accomplished, we [conclude the chapter with this problem, and the section with this chapter, and the book with the section]. Were it to be found pleasing to the illustrious intellect ... of that noble personage, fate will have assisted the success of the yearnings and the attainment of the desires of this sincere and blameless supporter. And if due to a transgression of the pen or fault of expression or feebleness of meaning or discordance of import the book is deprived of the honor of finding favor, it is hoped from that fount of excellence and generosity and that source of goodly character, that he conceal it with the cloth of forgiveness, as pardoning such errors by such a source of generosity would require no excuse ....[31]

Shīrāzī then makes an allusion to the vicissitudes that faced him during the composition of the *Ikhtīyārāt* (again echoing his words in the *Nihāya*) before calling more blessings upon the dedicatee and concluding his work.[32]

Though, as we have seen, Shīrāzī uses some of the same tropes in the introduction to all three of his works (e.g., the hardships faced by the author during the composition of the work, and the confident affirmation of his success in advancing the frontiers of astronomy) two of the features of the *Ikhtīyārāt* are rather striking and should be pointed out. The first, of course, is the metaphor-laden opening, that has no parallel in the other two works, and is replete with astrological references, including the metaphor of God as a dice-player rolling his dice in the "azure bowl" with the stars represented as "thousands of crystal game-pieces in the twelve mansions of this kohl-darkened plot." It is through the influence of these stars, we are told, that the creator can "at times, set the token of the actions and of the appointed times for the creatures of the world moving gainfully in creation and existence" or accomplish the opposite. (These words indicate that the patron's "pursuits," alluded to in the *Ikhtīyārāt* (and likely the *Tuḥfa*, as well) were astrological and divinatory.)

The second feature has already been touched upon and is Shīrāzī's discussion of the difficulties facing Ptolemaic astronomy and the great effort that has been expended in emending his astronomical theories. As concise and cogent a description of a centuries-long *hay'a* tradition as this passage represents, its placement seems immediately following the literary introduction is striking, and indeed raises

---

[30] Shīrāzī, *Ikhtīyārāt-i Muẓaffarī*, Ayasofya MS 2575, 2r.; Ṭūsī, *Ḥall-i mushkilāt-i muʿīnīyah*, 2.

[31] Shīrāzī, *Ikhtīyārāt-i Muẓaffarī*, Ayasofya MS 2575, 275r.

[32] Shīrāzī, *Ikhtīyārāt-i Muẓaffarī*, Ayasofya MS 2575, 275r.

the question: Why didn't Shīrāzī include a similar statement in the introduction to the *Nihāya*, i.e. the text that is considered his seminal work?[33] Though this question can't be answered definitively, the fact that this forceful phrasing appears in the *Ikhtīyārāt* implies that Shīrāzī expected the readers of this work to be fully appreciative of the problems of *hay'a* and the considerable effort needed to arrive at acceptable theoretical formulations.

Despite the similarities seen so far between the *Ikhtīyārāt* and the *Tuḥfa*, Shīrāzī does not spell out whether or not preserving the "preferred method" involved omitting from the *Ikhtīyārāt* astronomical knowledge that would have been included in the *Nihāya.* In the upcoming discussion we will therefore have the opportunity to examine this in an effort to verify Shīrāzī's claims as to the genesis of his books. Based on the author's claims, however, of the three *hay'a* texts under consideration here, the *Nihāya* appears to have been the primary work, with the *Tuḥfa* focusing ostensibly on the accepted (or "preferred") astronomical theory, and the *Ikhtīyārāt* consisting of a rendition of the *Nihāya* in Persian.[34] In the most general sense, of course, the three works reflect Shīrāzī's goal of providing the reader with a theoretically sound description of the heavens.[35]

4.3 The Structural Outline of the Works in Question

The similarity of the outline of the *Nihāya* and *Tuḥfa* to Ṭūsī's *Tadhkira*, as far as the outlines of these works are concerned, was first pointed out by Livingston.[36] This is in keeping not only with the influence that the great Ṭūsī must have exerted upon his student, but also with the influence of Ṭūsī's *Tadhkira* on the subsequent history of astronomy in the Islamic world.[37] Ragep, who authored the modern edition of the *Tadhkira*, notes the popularity of this work, and the fact that it was the subject of numerous commentaries.[38] The table of contents for each of the three works by

[33]Recall that in the introduction to the *Nihāya* Shīrāzī claims to have compiled the best of the works of the ancients and the moderns in his book.

[34]Yet, despite his claims to the contrary Shīrāzī appears to propose a number of different models for Mercury in the *Tuḥfa* (Prof. Saliba, personal communication).

[35]Indeed, while reading the *Nihāya* and Shīrāzī's other two works listed in the study one can hear echoes of Ibn al-Haytham's purpose for the composition of his *Maqāla*, namely, the transmission of "that which we understand of these sciences in order to instruct him who wishes to arrive at its comprehension without investigating." Abū Ibn al-Haytham, *Ibn al-Haytham's On the Configuration of the World*, 55. It should be noted as with Ṭūsī's *Tadhkira*, the reader of Shīrāzī's works is generally referred to the *Almagest* for the mathematical proofs of the topic under discussion. The notable exceptions are discussions involving novel formulations such as the Ṭūsī couple. See Prof. Ragep's discussion in Ṭūsī, *Naṣīr al-Dīn al-Ṭūsī's Memoir*, 36.

[36]J. Livingston, "Nasir al-Din al-Tusi's al-Tadhkirah," *Centaurus* 17 (1973): 260–275.

[37]Ṭūsī, *Naṣīr al-Dīn al-Ṭūsī's Memoir*, 56.

[38]Ṭūsī, *Naṣīr al-Dīn al-Ṭūsī's Memoir*, 56. Ragep also notes the fact that structurally the *Tadkhkira* is based on the *hay'a* works of Kharaqī. Ṭūsī, *Naṣīr al-Dīn al-Ṭūsī's Memoir*, 36.

Shīrāzī has been listed in Appendix A. The table of contents for the *Tadhkira* appears in Ragep's edition of this work.[39] The similarities of the *Ikhtīyārāt* to the *Nihāya* and *Tuḥfa* and to Ṭūsī's *Tadhkira* can be seen at once. Indeed, the organizational scheme in each of these books appears to be identical: In each work we have four books, the first containing introductory material, the second containing the configuration of the heavens, the third the configuration of the Earth, the fourth on measuring the distances of celestial bodies.

As far as Shīrāzī's books some quick observations about the layout of the three and their relations with each other are listed below:

Book 1: The division of the first book into three chapters is the same in all three works.

Book 2: The arrangement of the *second* book of the *Ikhtīyārāt* follows that of the *Nihāya.* The chapters are arranged differently in the *Tuḥfa*, however. Chapter One of the *Nihāya* appears to contain material that forms chapters *2*, *3*, and *4* of the *Tuḥfa.*

Book 3: The *Ikhtīyārāt* and *Tuḥfa* are in agreement as far as the arrangement of the third book is concerned. The minor difference between this book as it appears in these two works and the third book of the *Nihāya* is in the ordering of the chapters. Using the numbering of the chapters in the *Nihāya*, these chapters are arranged as follows in the *Tuḥfa* and the *Ikhtīyārāt*: 1, 2, 3, 4, 5, 6, 7, 11, 8, 9, 10, 12, 13. So, in the two later works, the material from chapter 11 of the *Nihāya* appears prior to the material that was presented in chapter 8 of the *Nihāya.*

Book 4: The *Ikhtīyārāt* and *Tuḥfa* are in agreement as far as the arrangement of the fourth book is concerned. This book consists of three chapters in each of these books. In contrast the fourth book of the *Nihāya* contains ten chapters. The titles of chapters 1, 2, and 3 in *Tuḥfa* correspond roughly to chapters 1, 9, and 10 in the *Nihāya.*

Given the close correspondence of the *Tuḥfa* and the *Ikhtīyārāt* for Books 1, 3, and 4, it may well be that, except for the Book 2, the *Tuḥfa*, was modeled on the *Ikhtīyārāt* (rather than on the *Nihāya*). It is Book 2, however, that contains the celestial models that are the specific subjects of this study, and in this chapter the *Ikhtīyārāt* is more similarly organized to the *Nihāya*, with the *Tuḥfa* deviating somewhat from its two predecessors. Shīrāzī himself was well aware of the importance of Book 2 for in the *Nihāya* he refers to it as the "main part of the book."[40]

---

[39] Ṭūsī, *Naṣīr al-Dīn al-Ṭūsī's Memoir*, x–xiii.

[40] Shīrāzī, *Nihāyat al-idrāk fī dirāyat al-aflāk*, Köprülü MS 957, 98r., and *Nihāyat al-idrāk fī dirāyat al-aflāk*, Köprülü MS 956, 83r. "And the true [account] in the resolution of the problem of Mercury rests upon the visualizing of its orbs in the manner preferred by us. We will thus describe first the orbs of the other planets in the manner in which these [are commonly accepted], indicating that which is preferred by us within it, we will then follow this at the end with the solution of Mercury and some of what remains from what we have promised to cover, then concluding the

When discussing the orbs of the planets in this chapter Ṭūsī follows a standard scheme by first discussing relevant observational data, and then presenting the number and alignment of the orbs, followed by the motions of these orbs, and the anomalies associated with the motions.[41] The outline of Shīrāzī's chapter on the superior planets in each of his *hay'a* works appears in Appendix B. As can be seen this scheme of Ṭūsī's is present, with some modifications, in all three of Shīrāzī's books. In the chapter on the superior planets, as these appear in Appendix B, perhaps the most significant difference is seen in a set of Shīrāzī's commentaries following the discussion of the planetary anomalies in the *Nihāya* and the *Ikhtīyārāt*. As far as the *Tuḥfa* a good deal of this material is omitted outright. We will look at Shīrāzī's commentaries following the discussion of the planetary anomalies in some detail in Chap. 5. The text of the first section of the chapter on the superior planets is reproduced in Appendix C, to provide the opportunity for a side by side comparison of the *Tadhkira*, the *Nihāya* and the *Ikhtīyārāt*. As can be seen there are many instances were the texts parallel each other; but there are as well many places, as well, in which they diverge.

4.4 Chronology Revisited

As was mentioned in the beginning of this chapter, the relative chronology of the *Nihāya* and the *Ikhtīyārāt* is complicated somewhat by a reference to the *Ikhtīyārāt* in the text of the *Nihāya* (which was completed prior to the *Ikhtīyārāt*). This occurs in Shīrāzī's discussion of the equant in the chapter on the superior planets. Commenting on the issue of the prosneusis point for the Moon Shīrāzī writes: "And its true cause is uniformity of motion, since for every sphere, the center of which is moving about a point with uniform motion, there exists a diameter that is aligned with this point, regardless of whether this point is at the center of the orbit of the sphere's center or not. And we have explained this in detail in the *Ikhtīyārāt-i Muẓaffarī*, to which it is incumbent upon you to pay heed, should you wish to be informed of it."[42] This statement, with its remarkable claim in regard to the alignment point, appears in the margin of the Köprülü 957, but has been incorporated into the body of the text in the later manuscript, Köprülü 956.[43]

It is not entirely clear how this cross-referencing could have come about. Given the fact that the two works were written in close succession, it appears as though some time after the completion of the *Nihāya* in the Fall of 1281 C.E. Shīrāzī

chapter which is in truth the main part of the book, by mentioning the configuration of the orbs of Venus and Mercury in our chosen method."

[41] Ṭūsī, *Naṣīr al-Dīn al-Ṭūsī's Memoir*, 416.

[42] Shīrāzī, *Nihāyat al-idrāk fī dirāyat al-aflāk*, Köprülü MS 957, 72r.

[43] Shīrāzī, *Nihāyat al-idrāk fī dirāyat al-aflāk*, Köprülü MS 956, 58v.

(who was obviously still in the process of revising his astronomical theory) decided to clarify his presentation regarding the prosneusis point for the Moon (which is one of the features of the Ptolemaic lunar model discussed in section V.5 of the *Almagest*). This emendation appears as a marginal note in MS Köprülü 957, which was copied at the end of the summer of 1282 C.E. referencing the work that he had since finished, the *Ikhtīyārāt.* In doing this Shīrāzī was revising a work that was purportedly to have contained his mature thoughts *hay'a.* Indeed, he appears to have liked his work in the *Ikhtīyārāt* well enough as to notify the reader about a fuller treatment of one of the fine points of his astronomical theory as provided in the *Ikhtīyārāt*, without bothering to provide an Arabic translation in the *Nihāya.*

There are many, more extensive, emendations appearing in MS *Nihāya* Köprülü 957. A considerable number of these are different from the marginal note we have just seen, in that they represent revisions (rather than referencing a fuller account of an issue that has been presented elsewhere), i.e., they are meant to replace what Shīrāzī had originally written in the *Nihāya.*

To add further richness to the relationship between the *Nihāya* and its companion work, the *Ikhtīyārāt*, it is worth noting here that a good deal of the revised text – that has been crossed out in the main text and replaced with comments in the margins of the text – survives in its original state in the *Ikhtīyārāt* (see Appendix E). This has several implications. The first is that the *Nihāya* was clearly the text in which Shīrāzī wanted to capture his changing thoughts on what were actively evolving models. The *Ikhtīyārāt*, in contrast, appears to have been left largely unchanged relative to its original edition (dating from 4 months after the completion of the *Nihāya* in November 1281). The second implication involves MS Köprülü 957. It is clear that this manuscript is one that belonged to the author and one in which his revisions were entered long after the official completion date for the work.

Based on what we have seen so far the chronological ordering of the works in question is therefore: *Nihāya* (Nov. 1281) – > *Ikhtīyārāt* (Feb. 1282) – > *Nihāya* MS Köprülü 957 (Aug. 1282) – > Emendations to MS Köprülü 957 (probably Aug. 1282, but certainly before Dec. 1284) – > MS Köprülü 956 (Dec. 1284) – > *Tuḥfa* (Aug. 1285). While a complete study of these works will likely uncover other manuscript traditions, this scheme should provide a fair description of the Shīrāzī's works on *hay'a* during the period 1281–1285 C.E.

4.5 The Hypotheses

Chapter Five of the second book of the *Tadhkira* is entitled "On basing some of the apparently irregular motions upon models that bring about their uniformity."[44] The Arabic word *aṣl* (plural *uṣūl*), translated above as "model" by Ragep, is most

[44]Ṭūsī, *Naṣīr al-Dīn al-Ṭūsī's Memoir*, 130.

generally translated as principle or axiom. Morrison translates the word *aṣl* (as this appears in Shīrāzī's work) as hypothesis, thus evoking the origin of this word in the *Almagest*. In this work I have followed Morrison's choice, but note here that to Ptolemy the word hypothesis would have meant a physical device used to mimic or model the motion of the celestial bodies.[45] In this chapter Ṭūsī presents his discussion of motions via epicycles and eccentric orbs (following the *Almagest* III.3), demonstrating – among other things – the well-known equivalence of eccentric motion to a motion composed of two circular motions with a shared angular motion, a formulation generally ascribed to Apollonius.[46] Not surprisingly, given Shīrāzī's debt to Ṭūsī's *Tadhkira*, each of Shīrāzī's three works on *hay'a* has an analogous *uṣūl* chapter, though the contents of these three chapters are somewhat varied.

The corresponding chapter in the *Tuḥfa*, entitled "On the Ascription of Apparently Irregular Motions Known Through Observation to Hypotheses [i.e., models] That Entail the Possibility of Their Arising from Orbs," has been translated and edited by Morrison.[47] Morrison's article includes a discussion of how Shīrāzī's conception of these *uṣūl* was different than that of earlier astronomers and especially that of Ṭūsī, who, unlike Shīrāzī, did not include formulations such as his own "Ṭūsī Couple" in his chapter on the "models/hypotheses." Ṭūsī discusses the "Ṭūsī Couple," in both planar and spherical variations in a subsequent chapter entitled "An indication of the solution – of that which is amenable to being solved – of the difficulties referred to previously that arise from the aforementioned motions of the planets."[48]

Shīrāzī's conception of the *uṣūl* appears, therefore, to have been in some ways more general than Ṭūsī's. In Chapter 2.5 of the *Nihāya* Shīrāzī presents a list of nine "hypotheses" (including those of the eccentric and the epicycle, but also the "Ṭūsī Couple" and the formulation based on 'Urḍī's Lemma).[49] The list of the *uṣūl* as they appear in the *Nihāya*, together with corresponding material in the other two works has been reproduced in Appendix D. A comparison of the hypotheses between the *Nihāya* and the *Tuḥfa* (i.e., columns 2 and 3 in Appendix D) demonstrates that in the period between composing the two works Shīrāzī's thinking in regard to the hypotheses appears to have changed. We see for example, that the number of irregularities of motion which he uses the *uṣūl* to address is fewer by two in the *Tuḥfa* than in the *Nihāya*.[50] Perhaps the most striking difference in the two works is the omission of one of the hypotheses, used in the *Nihāya* to treat the problem

---

[45]Hence the appeal of "model," to denote both a mathematical formulation as well as a physical device. See Ṭūsī, *Naṣīr al-Dīn al-Ṭūsī's Memoir*, 23; Swerdlow, *Mathematical Astronomy in Copernicus's De Revolutionibus*, 40.

[46]Ptolemy, *The Almagest*, 141.

[47]Morrison, "Qutb al-Din al-Shirazi's Hypotheses for Celestial Motions."

[48]Ṭūsī, *Naṣīr al-Dīn al-Ṭūsī's Memoir*, 195.

[49]Shīrāzī, *Nihāyat al-idrāk fī dirāyat al-aflāk*, Köprülü MS 957, 41r-51r.; See also Morrison, "Qutb al-Din al-Shirazi's Hypotheses for Celestial Motions."

[50]Morrison, "Qutb al-Din al-Shirazi's Hypotheses for Celestial Motions," 40.

of the equant in the superior planets. We will have occasion to revisit shortly this hypothesis and a companion hypothesis that is present in the *Nihāya* but that was omitted by Shīrāzī in the *Tuḥfa*, as well.

A sense of how Shīrāzī viewed these hypotheses can be seen in his presentation in the *Nihāya*, in which each hypothesis refers to an irregularity of motion for which it represents a solution (for example the epicycle and deferent can be used to account for retrograde motion). If the hypotheses themselves introduced undesirable features, these could be addressed by relying on other hypotheses, as can be seen in the following section from the *Ikhtīyārāt*: "[We insert a] maintainer orb between the [planetary] epicycle and the dirigent, [so that this maintainer is] centered on the center of the epicycle and its motion is equal to that of the dirigent [in magnitude but] in the opposite direction so that the motion of the dirigent is decoupled from the motion of the epicycle, and the motion of the epicycle remains simple and [that it does not include a contribution from other motions]."[51]

Also worth noting in regard to the "hypotheses" is the unusual features of the chapter on the *uṣūl* in the *Ikhtīyārāt.* As can be seen in Appendix D, the material in the *Ikhtīyārāt* is organized differently from the other two works. For one thing, many of the *uṣūl* themselves are scattered in the various chapters of Book 2. (In addition to Chapter 5, the *uṣūl* appear in Chapters 7, 8, and 9 of the *Ikhtīyārāt*).

A brief look at the introduction of the *Ikhtīyārāt* chapter on the *uṣūl* in Book 2 reveals that, unlike the analogous chapters in the other two works, it is formally divided into four sections: first, an "explication of the reason for fastness and slowness," second, an "explication of the reason for retrograde, station, and direct motion," third, on "the manner of imagining the corporeal orbs, in two and three dimensions," fourth on the "generally accepted configuration of the orbs, and a [subtle] indication of the issues facing it."[52] The first two sections cover much the same material at the beginning of the *uṣūl* chapters in the other two works. As far as the third section, i.e., the discussion of "the manner of imagining the corporeal orbs," related material also appears in both of the corresponding chapters from the *Nihāya* and the *Tuḥfa*.[53] The fourth section of this chapter, however (namely that of the "generally accepted configuration of the orbs"), does not appear in the corresponding chapter on the "hypotheses" in the other two works. As we will see in the discussion of the superior planets, material related to this section appears the *Nihāya* chapter on the superior planets (i.e., Book 2, Chapter 8) and appears to have

[51]Here, in addition to its desired effect of moving the center of the epicycle along a desired path, the motion of the dirigent (or encompasser) orb has resulted in a rotation of the enclosed epicycle. Shīrāzī relies on another hypothesis consisting of a single orb (i.e., the maintainer) to counter this undesired rotation.

[52]Shīrāzī, *Ikhtīyārāt-i Muẓaffarī*, Ayasofya MS 2575, 67v.

[53]Shīrāzī, *Nihāyat al-idrāk fī dirāyat al-aflāk*, Köprülü MS 956, 36v.; *al-Tuḥfa al-shāhīya*, BN Arabe MS 2516, 29r.

been omitted outright from the *Tuḥfa.* The location of this material in the *Nihāya* provides an important clue as to Shīrāzī's conception of this work, as we will see in the upcoming sections of this chapter.[54]

4.6 On the Moon

In his discussion of the orbs of the Sun, the Moon, and the superior and inferior planets, Shīrāzī's fixed presentational scheme is inherited from Ṭūsī. For each chapter, he starts by providing key observational data, and follows this by proposing the configuration of the orbs, their motion, and the observed *ikhtilāfāt* (i.e., non-uniformities or anomalies). In the following discussion, rather than fully describing Shīrāzī's models, I have reproduced his presentation of the orbs, to highlight basic features of his models and to draw several rather unexpected conclusions. Upon the complete study of these works, it is inevitable that further unexpected results will be brought to light. In the *Nihāya* chapter on the Moon Shīrāzī describes the configuration of its orbs as follows:

> The first orb is the **parecliptic** which is also called the *jauzahr* since upon its perimeter there is a point called the *jauzahr*, with its convex surface being [inwardly] tangent to the concave of Mercury and its concave surface tangent to the convex surface of the second of its orbs which is called the ***mā'il*** (inclined) which is a spherical body bound by two parallel surfaces the center of the two which is the center, as well, of the world, with its concave touching the center of convex of the sphere of fire of the four elements as is generally accepted and with its equator inclined relative to the pareclipic with a fixed inclination the limit of which based on what has been found through observation is five parts [i.e., degrees] and for this reason it has been called the inclined orb and its two poles are separated from the poles of the pareclipic in two reciprocal directions. The third orb is the **eccentric** in the thickness of the inclined orb in the aforementioned custom and its equator is in the plane of the equator of the inclined orb and with its two poles separated from the poles of the inclined orb in a single direction. The fourth orb is the **epicycle** in the thickness of the eccentric which carries it in a manner such that the distance of its center from the two poles of the eccentric is a single distance and the Moon is affixed within the epicycle in a manner such that its surface touches the surface of the epicycle at a point shared between the two and it accompanies its equator [i.e., that of the epicycle] which is the circle resulting from its [i.e., the Moon's] surface in the thickness of the epicycle.[55]

[54]It should be noted that *Ikhtīyārāt* exhibits a notable preoccupation with the "Conjectural Hypothesis", the "Deductive Hypothesis," and the "Innovative Hypothesis," or the *ḥadsī, istinbāṭī,* and *ibdāʿī* . These adjectives are all based on Arabic verbal nouns that have been transformed into adjectives in a practice that is common in the Persian-speaking world. In our discussion on the chapter on the superior planets we will be able to shed light on the meaning of the first two. An explication of the *ibdāʿī* awaits future studies.

[55]Shīrāzī, *Nihāyat al-idrāk fī dirāyat al-aflāk,* Köprülü MS 956, 44v. See Ṭūsī, Naṣīr al-Dīn al-Ṭūsī's Memoir, 150.

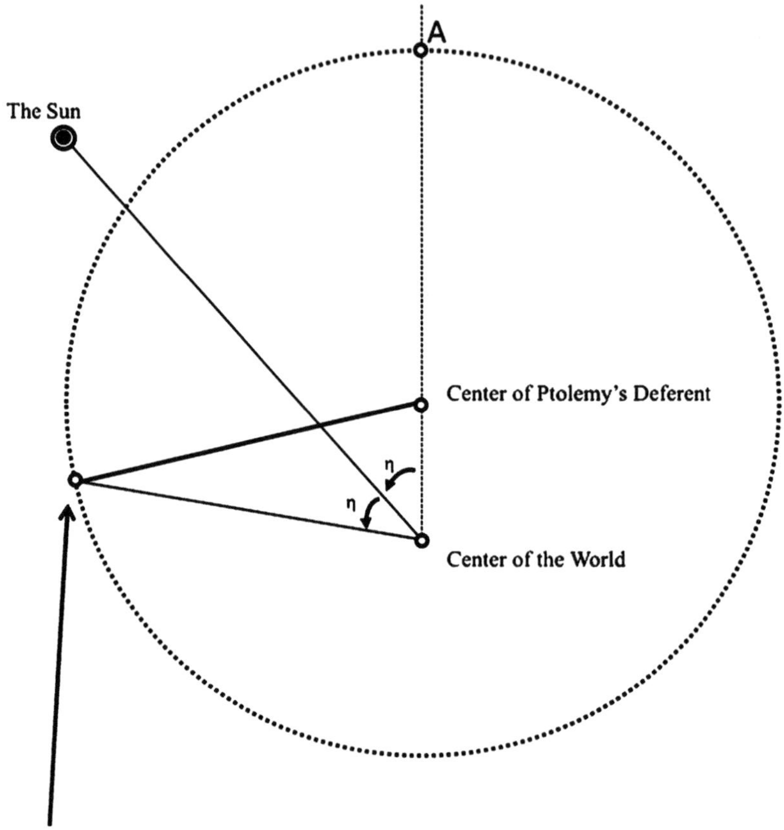

Fig. 4.1 The Model for the Moon in the *Nihāya* (First Model)

A schematic for this model appears in Fig. 4.1.[56] As has been noted by Saliba, a comparison of the model presented in this discussion with Ṭūsī's lunar model as presented in the *Tadhkira* chapter on the Moon indicates that the two are the same.[57] Shīrāzī subsequently discusses the issue of the motion of the deferent describing equal arcs in equal times about a point that is not its center (the well-known equant), as well as the *prosneusis* point.[58] In his discussion Shīrāzī invokes the Ṭūsī couple as well as providing a paraphrase of a section of the lunar model of al-'Urḍī from

[56]See Ṭūsī, Naṣīr al-Dīn al-Ṭūsī's Memoir, 160. The Moon's elongation is marked as η.

[57]Saliba, "Arabic Planetary Theories after the 11th Century AD," 97. Indeed the language of the two works bears a close affinity – again underscoring the debt of Shīrāzī's work to Ṭūsī. See Ṭūsī, *Naṣīr al-Dīn al-Ṭūsī's Memoir*, 149–150. This model ultimately derives from the *Almagest*. See also Saliba, "Arabic Planetary Theories after the 11th Century AD," 93.

[58]Saliba, "Arabic Planetary Theories after the 11th Century AD," 97.

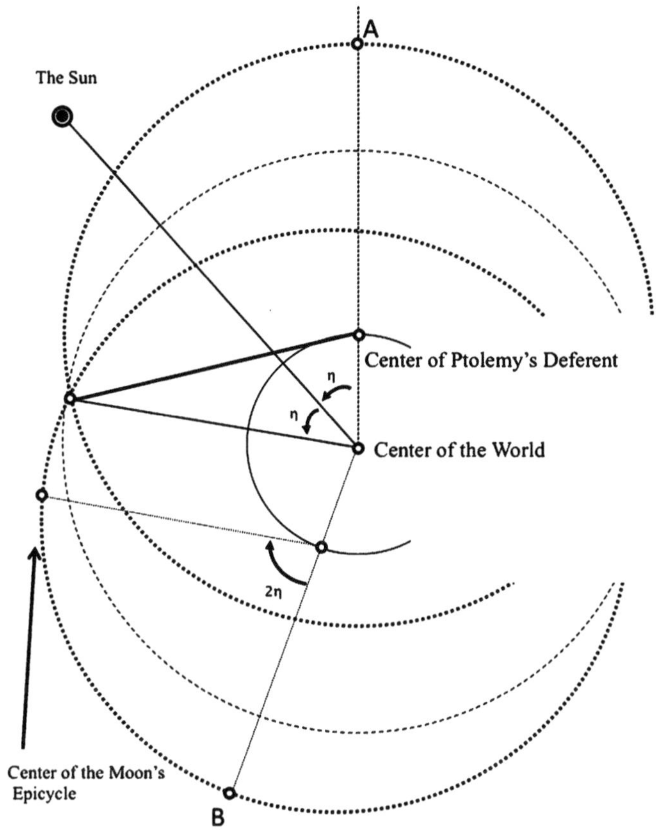

Fig. 4.2 'Urḍī's Model for the Moon in *kitāb al-hay' a*

his *Kitāb al-hay'a*. The lunar model of al-'Urḍī relies on the same number of orbs, but specifies a different configuration and different angular velocities for these orbs in order to treat the issue of the equant (Fig. 4.2).[59] Based on this information Saliba concludes his discussion of Shīrāzī's lunar model in the *Nihāya* by positing 'Urḍī's lunar model as Shīrāzī's "preferred" model in this work.[60]

By the time he authored the *Tuḥfa*, however, Shīrāzī presented a lunar model different from both that of Ṭūsī and al-'Urḍī. This model was formulated by Shīrāzī himself,[61] and it addresses the issue of the equant by using two rotating spheres – an eccentric deferent, with eccentricity half that of the Ptolemaic deferent, and the epicycle with a radius equal to the eccentricity – both rotating in the same sense.

[59]Saliba, "Arabic Planetary Theories after the 11th Century AD," 91.

[60]Saliba, "Arabic Planetary Theories after the 11th Century AD," 98.

[61]Saliba, "Arabic Planetary Theories after the 11th Century AD," 98–100.

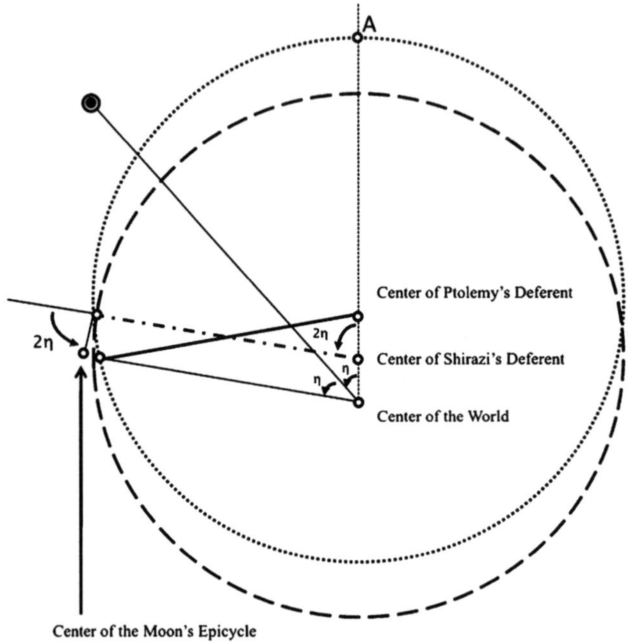

Fig. 4.3 The Model for the Moon in the *Nihāya* (Second Model)

Though this model is different from 'Urḍī's lunar model as it appears in the *Kitāb al-hay'a*, it relies on the mathematical formulation known as 'Urḍī's lemma (see Fig. 4.3), which al-'Urḍī had used for his model of the superior planets.[62]

Since Shīrāzī describes 'Urḍī's lemma in detail in the *Nihāya*, it is a bit surprising not to see him use this formulation in the first set of his models that one encounters in the *Nihāya*. Though, at first glance, one could ascribe this to the fact that Shīrāzī only realized how to adapt this formulation to his lunar model when he was completing the *Tuḥfa* (i.e., after completing the *Nihāya*).

To get a better sense of Shīrāzī's view in regard to the use of 'Urḍī's lemma for his lunar model it is useful to look at the *Ikhtīyārāt*, the companion work to the *Nihāya* that was written a few months later. A look at Shīrāzī lunar model in the *Ikhtīyārāt* reveals, quite unexpectedly, that this model is an entirely different model than the one we have seen in the *Nihāya*. For one thing the lunar model in the *Ikhtīyārāt* relies on six orbs versus the four that appear in the *Nihāya*! The first three orbs of this new *Ikhtīyārāt* model, the parecliptic, inclined, and deferent orbs, are defined as in the lunar model that is outlined in the *Nihāya* (which in turn is quoted from Ṭūsī's *Tadhkira*). Two additional orbs are inserted at this point; one enclosing the other, with the sixth orb, i.e., the Moon's epicycle nestled in the innermost.

---

[62] Saliba, "Arabic Planetary Theories after the 11th Century AD," 98–100.

The orbs that make their appearance specifically for the *Ikhtīyārāt* are described as follows:

> And the fourth orb is the **encompasser**, in the thickness of the eccentric in the accepted manner with its equator in the plane of the equator of the eccentric and its axis perpendicular to the plane of the equator of the inclined orb, and the fifth orb is the **maintainer** nested within the encompasser with its convex surface touching the concave surface [of the encompasser] at a single point and with its center separated from the center of the encompasser by the amount of the separation of the center of the corporeal (*mujassam*) deferent from the center of the world, and with its equator in the plane of the equator of the encompasser and its two poles [separated] in the same the direction from the two poles of the encompasser and its axis parallel to the axis of the encompasser.[63]

In the *Ikhtīyārāt* section immediately preceding the fragment listed above Shīrāzī contrasts an "imaginary deferent" (*mutawahham*) to the "corporeal deferent" (*mujassam*) seen above, stating in regard to the latter "and the third orb is the orb of the eccentric [deferent] in the thickness of the inclined orb as is well-known ... with its distance from the center of the world equal to half of that which is generally accepted ..., since that is the distance of the center of the imaginary deferent from the center of the world, not that of the corporeal deferent."[64] This adjustment, the establishment of a new center for the deferent (which in turn carries with it the encompasser, the maintainer, and the lunar epicycles), is just what is needed to allow for the application of 'Urḍī's Lemma in the configuration of the Moon.[65]

What is critical to the implementation of 'Urḍī's Lemma, however, is not only the location of the center of one of the newly added epicycles, i.e., the emcompasser, but the sense of rotation of the encompasser relative to the rotation of the deferent. 'Urḍī's Lemma requires these two rotations to have the same direction or sense. Shīrāzī indicates the motions of both the deferent and the encompasser orbs as being sequential, i.e., in the direction of the order of the signs of the zodiac. As a result, his proposed model contains an application of 'Urḍī's Lemma, which allows the center of the epicycle to move along the orbit predicted by Ptolemy's lunar model (or, rather, very close to it), while at the same time avoiding the physical contradiction of Ptolemy's model, which has the epicycle center move upon a deferent while "rotating" uniformly about a point distinct from the center of the deferent (i.e. the equant).

The question raised by Shīrāzī's lunar model in the *Ikhtīyārāt* is why he didn't use 'Urdī's Lemma in the *Nihāya*, completed four short months before the *Ikhtīyārāt.* The fact that the proof for 'Urḍī's Lemma is presented in the *Nihāya* in

---

[63] Shīrāzī, *Ikhtīyārāt-i Muẓaffarī*, Ayasofya MS 2575, 84r.

[64] Shīrāzī, *Ikhtīyārāt-i Muẓaffarī*, Ayasofya MS 2575, 84r.

[65] The *encompasser* is the additional epicycle that Shīrāzī relies upon to implement 'Urḍī's Lemma. He describes the *maintainer* orb as being with concentric with the lunar epicycle. This orb is used to rotate the lunar epicycle about its center (thus affecting the alignment of the orb). It is, therefore, not relevant as far as the motion of the center of the lunar epicycle is concerned, and is not necessary for the implementation of 'Urḍī's Lemma. Shīrāzī, *Ikhtīyārāt-i Muẓaffarī*, Ayasofya MS 2575, 84r.

the chapter on the superior planets only deepens the mystery.[66] Why present 'Urḍī's Lemma in the *Nihāya* and not use it for the configuration of the Moon (as Shīrāzī ended up doing, in both the *Tuḥfa* and the *Ikhtīyārāt*)?

The mystery of the absence of 'Urḍī's Lemma in the model for the Moon is solved, however, upon viewing another section of the *Nihāya* deeper into the book. It is in the section on the planetary latitudes, after a lengthy discussion of the latitudes that Shīrāzī presents another lunar model.[67] This one resembles that of the *Ikhtīyārāt*, and the *Tuḥfa*, and relies on 'Urḍī's Lemma.[68] The only sensible explanation is that the earlier sections on the Moon in the *Nihāya* is devoted to the presentation of the "generally accepted models," much as we saw in the chapter on the hypotheses in the *Ikhtīyārāt.* It is only at the conclusion of the section on the latitudes (which appears later in the book) that Shīrāzī proposes his own lunar model which relies on his implementation of 'Urḍī's Lemma.[69]

The fact that Shīrāzī organizes the *Nihāya* in this strange fashion is rather puzzling. What it has meant for modern scholarship on Shīrāzī, is the assignment of one of Shīrāzī's original contributions to *hay'a* – his lunar model – to his later work, the *Tuḥfa.* That the importance of his model – which was already present in the two early works, the *Nihāya*, and the *Ikhtīyārāt* – was apparent to the author is clear, however, in his introductory statements in the *Nihāya* and the claim of producing solutions that "had not occurred to anyone prior to me ...."[70]

4.7 On the Superior Planets

A comparison of the number of orbs included for the superior planets as they appear in each of the corresponding chapters in the *Nihāya*, *Tuḥfa*, and *Ikhtīyārāt* indicates that we are, again, dealing with distinct models.[71] This can be seen by referring

[66]Shīrāzī, *Nihāyat al-idrāk fī dirāyat al-aflāk*, Köprülü MS 956, 59v.

[67]Shīrāzī, *Nihāyat al-idrāk fī dirāyat al-aflāk*, Köprülü MS 956, 77r.

[68]The location of this section is vaguely reminiscent of Ṭūsī's placement of the "Ṭūsī Couple" in the *Tadhkira*. Ṭūsī, however, discusses the "Ṭūsī Couple" couple, in a new section, as we saw. It is in this section that Ṭūsī proposes a new model of the Moon relying on his new mathematical formulation.

[69]The lunar model in the *Tuḥfa* is similar to the other two. The primary difference is that Shīrāzī's thinking with respect to the question of alignments had apparently changed and he no longer saw a need for a "maintainer orb."

[70]Shīrāzī, *Nihāyat al-idrāk fī dirāyat al-aflāk*, Köprülü MS 957, 197r. See note 13.

[71]It is worth noting that in the *Nihāya* chapter on the upper planets the planet Venus is treated together with the upper planets Saturn, Jupiter, and Mars. In this Shīrāzī is following Ṭūsī's who in turn is following Ptolemy. This grouping obliges Shīrāzī to insert numerous comments pertaining specifically to Venus in his chapter on the superior planets. In the *Tuḥfa* and the Ikhtīyārāt however Venus is treated in the same chapter as Mercury. On Venus's similarities to the upper planets Ptolemy writes in *the Almagest* X.6: "Such, then, were the methods which we successfully used for these two planets Mercury and Venus, to establish the hypotheses and demonstrate [the sizes of] the anomalies. For the other three, Mars, Jupiter and Saturn, the hypothesis which we find for their

to Appendix B, section 2, "The Orbs." Shīrāzī lists three orbs for each of the superior planets in the *Nihāya* (i.e., the parecliptic, the eccentric deferent, and the epicycle of the planet), whereas in the *Ikhtīyārāt* he lists six (i.e., the parecliptic, the eccentric, the encompasser, the dirigent, the maintainer, and the planetary epicycle). In the *Tuḥfa* the list has shrunk down to five (i.e., the parecliptic, the eccentric deferent, the encompasser, the inclined orb, the epicycle of the planet). As with Shīrāzī's treatment of the Moon, much of the material that appears in the *Nihāya* chapter ostensibly dealing with the superior planets faithfully presents what Ṭūsī has described in the *Tadhkira*:

> And so they established three orbs and three motions for each of the four [planets]. The first orb is the **parecliptic**. For Saturn, its convex surface is contiguous with the concave surface of the eighth orb, and its concave surface is contiguous with the convex surface of Jupiter's parecliptic. The concave surface of Jupiter's parecliptic is contiguous with the convex surface of Mars's parecliptic. The concave surface of Mars's parecliptic is contiguous with the convex surface of the Sun's parecliptic. The convex surface of Venus's parecliptic is contiguous with the concave surface of the Sun's parecliptic, while its concave surface is contiguous with the convex surface of Mercury's parecliptic. And the second is the **eccentric deferent** [that carries] the epicycle. It is located in the thickness of the eccentric and for this reason it is called the deferent and the planets are embedded in the **epicycle**.[72]

Indeed a comparison with the corresponding lines in the *Tadhkira* indicates that in this case Shīrāzī is quoting verbatim from Ṭūsī's book.[73] In the *Ikhtīyārāt*, however, the three additional orbs that appear are introduced as follows:

> Due to the situation of these planets they demonstrated [the existence of] three orbs. However, based likewise on empirical observations, that will be described in their place, [it is known that] the uniformity of motion is about the equant point and the alignment of the mean apogee is also with respect to the same point, and the inclination of the [epicylic] diameter passing through the apogee and perigee [of the epicycle] relative to the inclined plane occurs in a specific manner, and none of [these phenomena] can result from the three [aforementioned] orbs, so we were compelled to add three orbs for each of these planets so that the sum was six orbs and six motions, and so that these observations could be derived from the proper arrangement of these orbs.[74]

Here we see again Shīrāzī echoing his remarkable claim in regard to the coincidence of the alignment point and the equant for the Moon (first encountered in the quote from the *Nihāya* presented at the beginning of Sect. 4.4 of the current chapter). It appears as though part of Shīrāzī's scheme is centered on providing a solution that addresses the issue of the equant, as well as accounting for planetary latitudes. Ṭūsī had already attempted to treat both of these issues in the *Tadhkira*, the first with

motion is the same [for all] and like that established for the planet Venus, namely one in which the eccentre on which the epicycle center is always carried is described on a center which is the point bisecting the line joining the center of the ecliptic and the point about which the epicycle has its uniform motion." Ptolemy, *The Almagest*, 480.

[72]Shīrāzī, *Nihāyat al-idrāk fī dirāyat al-aflāk*, Köprülü MS 956, 44v.

[73]Ṭūsī, *Naṣīr al-Dīn al-Ṭūsī's Memoir*, 181.

[74]Shīrāzī, *Ikhtīyārāt-i Muẓaffarī*, Ayasofya MS 2575, 106r.

the inclusion of a planar Ṭūsī couple mechanism and the second with the spherical Ṭūsī-couple.[75] Shīrāzī's presentation in the *Ikhtīyārāt* is a clear indication that he considered his own treatment to be superior to that of his teacher. Immediately after enumerating the orbs and the motions for the superior planets, Shīrāzī proceeds to describe the orbs themselves. The outermost orb, the parecliptic, is the same as it was in the *Nihāya*.[76] In regard to the next orbs in the sequence, Shīrāzī writes:

> The second is the **deferent** in the thickness of the parecliptic as is well-known and they call it the deferent, not because they have imagined that its equator is what conveys the center of the epicycle, for this is not true as will be shown, rather this is because the center of the epicycle is as one of the parts of the deferent.... And the third is the **encompasser**, with its center on the equator of the deferent and its convex touching the convex and concave of the deferent at two points and its equator intersecting the equator of the deferent by the fixed amount of the maximum inclination of the apogee of that planet. And the fourth is the **dirigent** centered on the center of the encompasser and enclosed with it, yet with its equator eternally in the plane of the equator of the deferent, and its axis intersecting the axis [of the encompasser] at [their centers]. And the fifth is the **maintainer** enclosed within the encompasser in a way such that its equator is in the plane of the equator of the encompasser and its center separated from the encompgsser's center by the amount of the distance between ... the center of the world and the center of the deferent... And sixth the **epicycle** orb within the maintainer such that their centers and equator and diameter are in agreement and with its equator never departing from the equator of the maintainer and the planet upon the epicycle moving along [the epicycle's] equator.[77]

The addition of three new orbs, each with a specified inclination presents a rather more complicated picture in latitude, affirming Shīrāzī's preoccupation with presenting a coherent description of latitude in his newly proposed model. Here we see Shīrāzī, as we did earlier, contrasting the (commonly assumed, and to Shīrāzī, false) *mutawahham* deferent (with its associated issues of non-uniformity of motion) and the true *mujassam* deferent.

Given the *mutawahham*, and *mujassam* nomenclature that we have already encountered in Shīrāzī's discussion of the Moon, and given the fact that, in his modeling of the superior planets, al-'Urḍī himself used the mathematical formulation that today bears his name,[78] it is natural to assume that Shīrāzī's arrangement for the superior planets in the *Ikhtīyārāt* would be based again on an implementation of 'Urḍī's Lemma. Further reading of this section in the *Ikhtīyārāt* conveys yet another surprise, however. In a description of the motions of the orbs he has decreed for the superior planets Shīrāzī writes:

---

[75] Ṭūsī, *Naṣīr al-Dīn al-Ṭūsī's Memoir*, 208 and 218. Part of Shīrāzī's concern is clearly the treatment of the prosneusis point for the Moon; again signaling his rejection of Ṭūsī's approach. See Ṭūsī, *Naṣīr al-Dīn al-Ṭūsī's Memoir*, 220.

[76] Rather than following the scheme in the *Nihāya*, Shīrāzī follows 'Urḍī's scheme by placing Venus's convex adjacent to the concave of Mars, as he does indeed for the *Tuḥfa*, when he describes the order of the nested orbs. See note 71.

[77] Shīrāzī, *Ikhtīyārāt-i Muẓaffarī*, Ayasofya MS 2575, 106r.

[78] Saliba, "Arabic Planetary Theories after the 11th Century AD," 104–108.

> And third, the motion of encompasser equal to the motion of its own center meaning the motion of the eccentric of that planet such that in the manner that in the upper half it is against the motion of the eccentric meaning counter-sequential. And fourth is the motion of the dirigent as the motion of the encompasser exactly, in both direction and measure. And fifth the motion of the maintainer which is twice the motion of the encompasser and sequential in the upper half. So due to the equivalence of the motion of the encompasser and the deferent [in magnitude] and the opposition [in direction] as well as the fact that we assumed the distance between the center of the epicycle and that of the encompasser is equal to the difference [between the center of the world and the center of the deferent] what results from the [motion] of the center of the epicycle through the compounded motion of these two [i.e., the encompasser and the deferent] is an orbit equal to the equator of the deferent, as was described in the Conjectural Principle at the conclusion of the Chapter on the Moon.[79]

The direction indicated by Shīrāzī for the rotation of the encompasser orb here is opposite what one would expect for an implementation of 'Urḍī's Lemma. With the encompasser orb rotating in the opposite sense of its deferent, the resulting configuration resembles a description of Apollonius's Theorem, in which the motion caused by an eccentric deferent can be shown to be equivalent to the motion caused by the combination of a concentric deferent and an epicycle. This "hypothesis" appears in Shīrāzī's own list in the *Nihāya* as number 3 (see Appendix D). Shīrāzī is within reason in stating that this hypothesis could be configured to cause the motion of the epicyclic center to be uniform about an "equant" point. However, if one were to choose the direction of motion as Shīrāzī describes in this passage, the resulting trajectory would deviate grossly from the expected one, i.e., the Ptolemaic deferent which was determined by observation and had to be maintained.[80] This unexpected feature of Shīrāzī's treatment of the superior planets in the *Ikhtīyārāt* was only discovered thanks to a careful reading of this work by Gamini.[81]

Could this reading of the *Ikhtīyārāt* text be in error? Could the desired sense of the rotation of the encompasser have been the opposite of what we have assumed it to be based on our reading of his text? Shīrāzī himself provides an additional clue by referencing the "Conjectural Principle" at the end of the chapter on the Moon. There he describes the motion of his encompasser orbs as "equal to the motion of their centers, meaning the motion of the Eccentric ..., and in the upper half this is in the direction opposite to that of the Eccentric,"[82] thus leaving no doubt that his intended direction of rotation for the encompasser is the one we have assumed.

The reason for Shīrāzī's choice of model is unclear, especially since near the end of the very same chapter in which the model for the superior planets is presented,

---

[79] Shīrāzī, *Ikhtīyārāt-i Muẓaffarī*, Ayasofya MS 2575, 108v.

[80] For a schematic of Apollonius's Theorem, see Figure 4.5. Shīrāzī refers to this as the Conjectural-Superior in the *Ikhtīyārāt*.

[81] Amir Mohammad Gamini, "The Planetary Models of Quṭb al-Dīn Shīrāzī in the *Ikhtīyārāt-i Muẓaffarī*," *Tarīkh-i 'ilm*, no. 8 (1388): 39–54. This finding was originally published by Mr. Gamini at the International Congress of History of Science, Budapest, 2009.

[82] Shīrāzī, *Ikhtīyārāt-i Muẓaffarī*, Ayasofya MS 2575, 102v.

we see one of the several descriptions of ʻUrḍī's Lemma in the *Ikhtīyārāt*.[83] At the conclusion of this discussion Shīrāzī comments upon the utility of this principle for treating issues related to the equant:

> And if one ponders this hypothesis (*aṣl*) it becomes apparent that the characteristic of this situation is such that the motion of a point that is moving by a compound motion is uniform about a point the distance of which relative to the center of the corporeal deferent is equal to the distance of the moving point relative to the center of the dirigent (*mudīr*).[84]

Here, Shīrāzī calls the additional epicycle which encases the epicycle of the planet (and which he previously referred to as the encompasser) the dirigent. The use of the term dirigent here is confusing at first, but it also helps clarify the labeling of ʻUrḍī's Lemma in both the *Nihāya* and the *Ikhtīyārāt* as the hypothesis of the dirigent and the maintainer.[85] Shīrāzī continues:

> And the situation of that point will be different depending on the situation of the epicycle. For if the center of the epicycle [i.e., point E in Fig. 4.4] is assumed at the lower half of its orbit [circle ESA] as the master of this principle has it, by necessity the uniformity of motion will be relative to a point above the center of the corporeal deferent [point D], and if it is assumed in the upper half [of circle ESA, i.e., at point A] then uniformity of motion will be relative to the point below the center of the corporeal deferent. And since for the Moon the desired uniformity was relative to a point below the imaginary deferent, we had no recourse but to set the center of the corporeal deferent below the center of the imaginary deferent so that which was desired would be achieved.[86]

That the location of the epicycle at the apogee of the deferent is a relevant parameter had been implicitly expressed by Ṭūsī in his discussion of the application of the Ṭūsī couple for celestial bodies other than the Moon.[87] Shīrāzī follows the same approach with regard to the configurations based on ʻUrḍī's Lemma as well as those based on Apollonius's theorem:

> And know that these two laws [sing. *ḥukm*] are only useful when the motion of the dirigent in the upper half is the same as the motion of the deferent. For if [the motion is in the opposite direction] both laws are inverted, such that in this reckoning if the center of the epicycle is assumed to be above the center of it [i.e., the dirigent] uniformity will be relative to a point above the center of the corporeal deferent and if it is assumed to be under [the center of the dirigent] then uniformity of motion will be relative to a point below [the center of the dirigent] then uniformity of motion will be relative to a point below [the center of the corporeal deferent]. And according to these [two last] schemes the center of the epicycle

---

[83]Shīrāzī, *Ikhtīyārāt-i Muẓaffarī*, Ayasofya MS 2575, 114r.

[84]Shīrāzī, *Ikhtīyārāt-i Muẓaffarī*, Ayasofya MS 2575, 114v.

[85]As we will see Shīrāzī uses as well the term Deductive Hypothesis for ʻUrḍī's Lemma. When referring to ʻUrḍī's Lemma in the configuration of the superior planets in his two earlier books he uses the term Dirigent to refer to a concentric orb encased by the Encompasser, with an axis of rotation that is tilted relative to that of the Encompasser. In this configuration the Dirigent merely appears to help Shīrāzī' with his bookkeeping of the latitude. He uses the label encompasser for the orb encasing the epicycle of the Moon in the ʻUrḍī picture.

[86]Shīrāzī, *Ikhtīyārāt-i Muẓaffarī*, Ayasofya MS 2575, 114v.

[87]Ṭūsī, *Naṣīr al-Dīn al-Ṭūsī's Memoir*, 446.

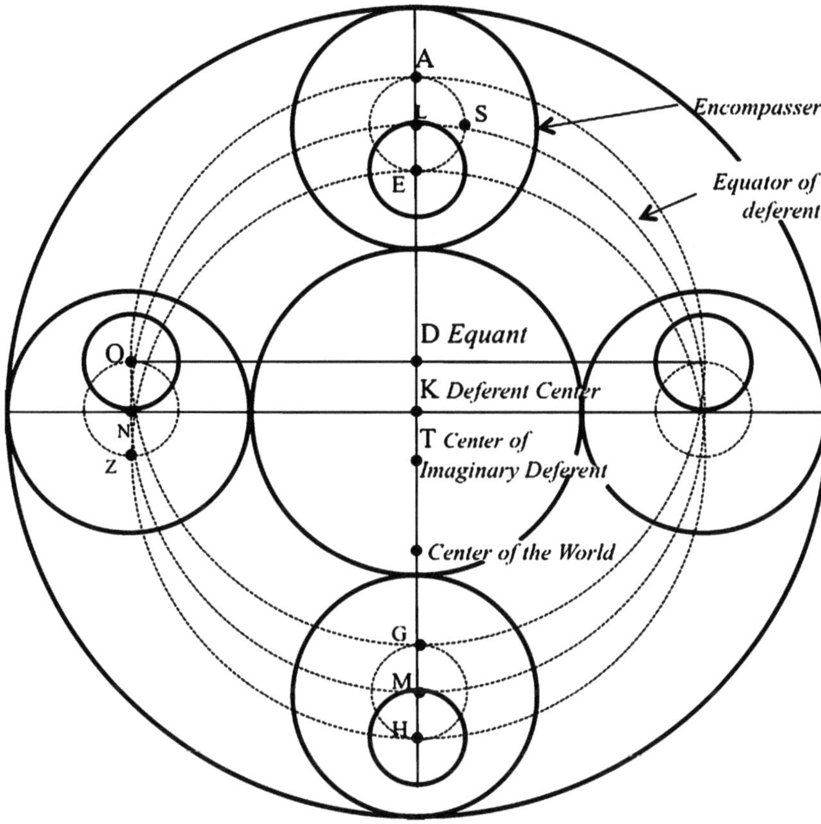

Fig. 4.4 Shīrāzī's presentation of the Deductive Hypothesis (based on 'Urḍī's Lemma) in the *Ikhtīyārāt* chapter on the superior planets

> traverses a [truly circular trajectory] such that the [point about which the motion is uniform] is at the center, unlike the case of the first two reckonings, since for them the center [i.e., the center of the secondary/small epicycle] does not traverse a truly circular trajectory.[88]

Here Shīrāzī is pointing out that the predicted trajectory for a body moving via a configuration based on 'Urḍī's Lemma experiences a minor deviation from a circular orbit (while implicitly recognizing as well that these deviations from a circular trajectory are negligible for the physical parameters of the Solar System). Could this deviation, then, be the reason for Shīrāzī's rejection of 'Urḍī's Lemma for the superior planets? We will return to this question in Chap. 5.

Shīrāzī labels "that division of the 'deductive' that requires a uniformity of motion relative to a point above the center of the embodied deferent" the "superior" and the other the "inferior." We are left with a four-part scheme involving 'Urḍī's

[88]Shīrāzī, *Ikhtīyārāt-i Muẓaffarī*, Ayasofya MS 2575, 114v.

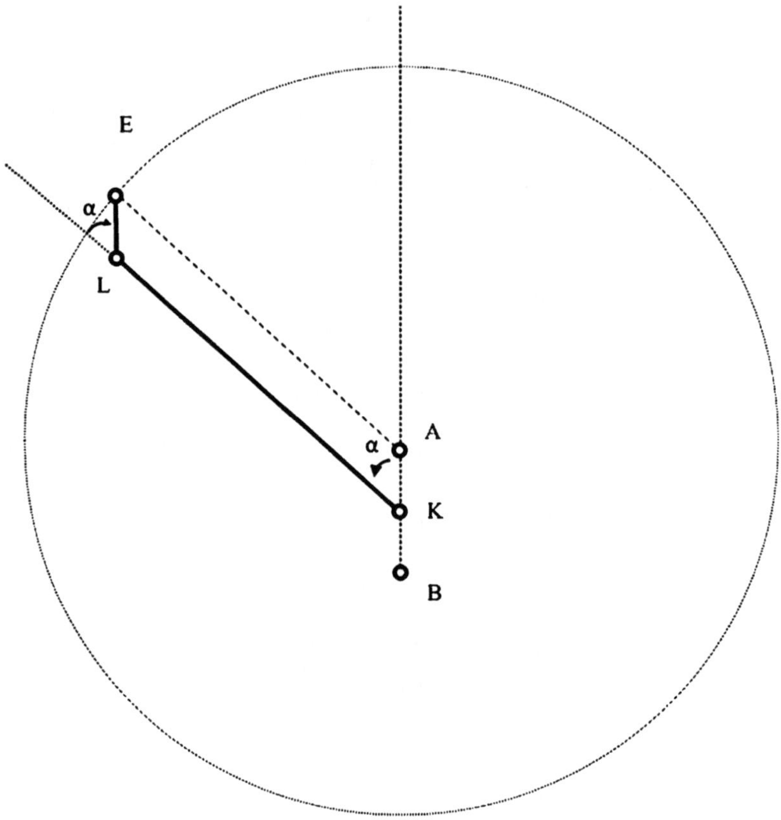

Fig. 4.5 The Conjectural-Superior Hypothesis

Lemma and Apollonius's Theorem. An illustration of these four "hypotheses," to which Shīrāzī refers to more than once in the *Nihāya*, but which only receive a full explanation in the *Ikhtīyārāt* appears in Figs. 4.5, 4.6, 4.7, and 4.8.

The Conjectural Hypotheses result from a deferent and an encompasser that are turning in the opposite sense (e.g., with the deferent turning counterclockwise and the encompasser turning counter-clockwise, or vice versa). Shīrāzī takes the commonly presented configuration of Apollonius's theorem, and calls this the Conjectural-Superior. In the Conjectural-Inferior configuration the deferent and the encompasser maintain the same sense of rotation as that of the Conjectural-Superior but the orientation of the encompasser at the apogee of the deferent is different: for this configuration the encompasser is rotated about its center L by 180 degrees relative to its analogous configuration in the Conjectural-Superior (see Figs. 4.5 and 4.6).

As can be seen in the figure, the Deductive Hypotheses are based on 'Urḍī's Lemma. As before, the Deductive-Inferior and Deductive-Superior are related by a

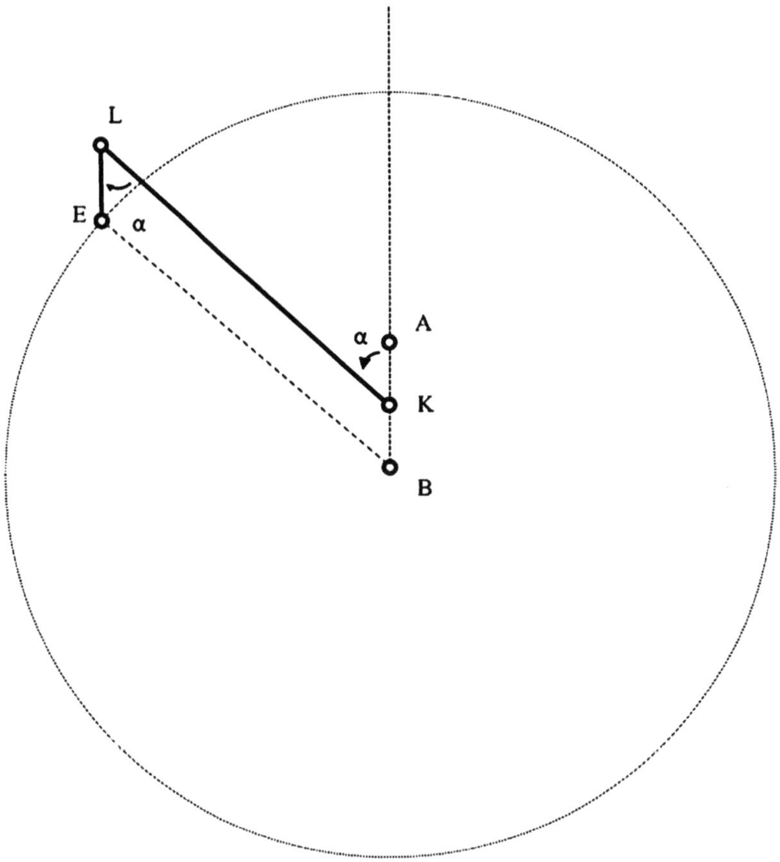

Fig. 4.6 The Conjectural-Inferior Hypothesis

simple rotation of the encompasser (about its center L) by 180 degrees (see Figs. 4.7 and 4.8). In both the Deductive-Superior and the Conjectural-Superior the motion of the center of the epicycle (marked E in Figs. 4.5 and 4.7) appears as though it is uniform relative to a point falling above the center of the deferent, i.e., point A in the figure. In the Deductive-Inferior and the Conjectural-Inferior, however, the motion of the center of the epicycle E is uniform relative to a point falling below the center of the deferent (i.e., point B in Figs. 4.6 and 4.8).

The Conjectural-Superior Hypothesis (Shīrāzī's choice for the superior planets in the *Ikhtīyārāt*) was one that he abandoned in the *Tuḥfa.* In that work Shīrāzī opts for a configuration based on the Deductive-Superior Hypothesis (which is how, in his *Ikhtīyārāt*, Shīrāzī refers to what we refer to as 'Urḍī's Lemma). While considering Shīrāzī's choice of model for the superior planets in these two works it is important to remember that he had successfully implemented 'Urḍī's Lemma in the lunar

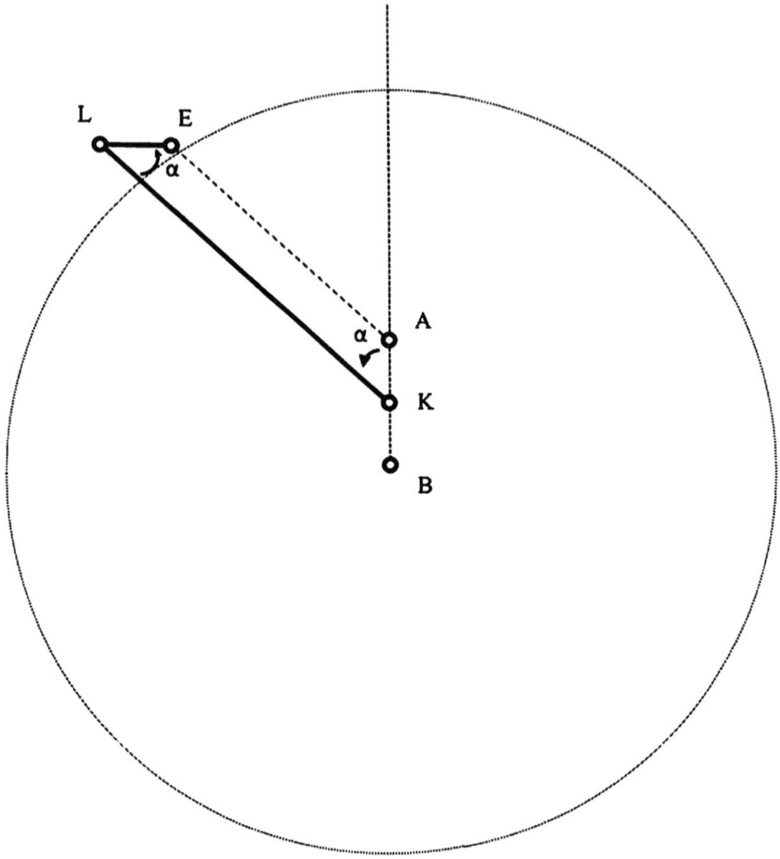

Fig. 4.7 The Deductive-Superior Hypothesis

model before he wrote the *Nihāya*.[89] Indeed, his choice of model for the superior planets here is especially surprising because Shīrāzī leaves no doubt that he was aware of 'Urḍī's models for the superior planets themselves, as he affirms by stating: "And since for [the superior planets] the desired uniformity of motion is relative to point above the center of the imaginary deferent, the master of this hypothesis had to assume that the center of the embodied deferent was above the imaginary deferent."[90] By "the master of this hypothesis" Shīrāzī is plainly referring to al-'Urḍī, and the configuration refers to 'Urḍī's configuration as this appears in his model for the superior planets.

[89]Shīrāzī, *Nihāyat al-idrāk fī dirāyat al-aflāk*, Köprülü MS 956, 77r. See also, Shīrāzī, *Ikhtīyārāt-i Muẓaffarī*, Ayasofya MS 2575, 115r.

[90]Shīrāzī, *Ikhtīyārāt-i Muẓaffarī*, Ayasofya MS 2575, 115r.

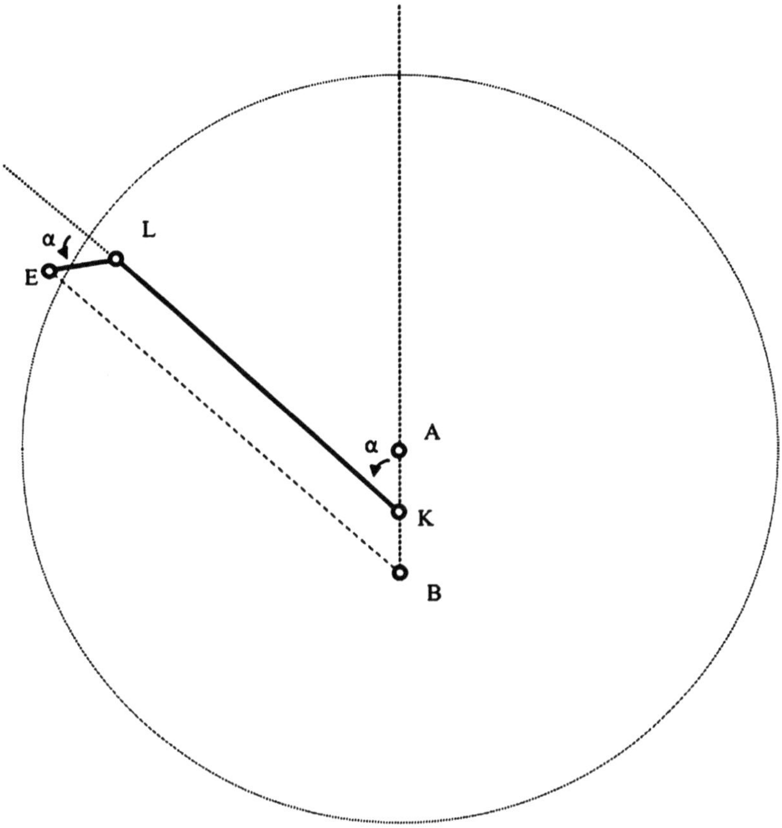

Fig. 4.8 The Deductive-Inferior Hypothesis

So much for the treatment of the superior planets in *Ikhtīyārāt*. Following the example of the Moon it is tempting to look at other sections within the *Nihāya* and to see what Shīrāzī has included there. Prior to an extensive search of this work however, it is easy to see that what Shīrāzī has included in the opening part of his chapter on the superior planets reproduces Ṭūsī's model in the chapter on the superior planets.[91] However, in order to answer the question "Did Shīrāzī propose an original model for the superior planets in the *Nihāya*?" we again need to look no further than the end of the chapter on the planetary latitudes, where, after having dispensed with his model of the Moon, Shīrāzī states:

> And as for the orbs of the superior planets each **includes six orbs, three of them the ecliptic and the deferent and the epicycle** as [accepted by all] as far as motions and the magnitude and directions of these and in their location except for the epicycle, **and the remaining three are those which we have added**. The first an **encompasser** orb and it is

[91] Shīrāzī, *Nihāyat al-idrāk fī dirāyat al-aflāk*, Köprülü MS 956, 55v.

> in the thickness of the epicycle with its equator always fixed to the [plane of the] equator of the deferent. And the second the **dirigent** in the thickness of the encompasser and with the same center as it, yet with its equator crossing the equator of the deferent and constantly inclined with respect to it, by as much as the inclination of that planet from the inclined orb, and its axis intersecting the axis of the encompasser at the center. And the third the **maintainer** enclosed by the dirigent, such that its equator is in the plane of the equator of the dirigent and its center separated from the center of the dirigent by the distance separating the center of the deferent of the planet from the center of the world, with this point [i.e., the center of the maintainer] lying upon the plane of the equator of the dirigent and with its axis parallel to the axis of the dirigent. And the **epicycle** is in the maintainer agreeing with it as far as equator and center and poles and axis at all times.[92]

Based on this fragment we see that Shīrāzī's proposed configuration for the superior planets in the *Nihāya* is virtually identical to that which he subsequently presented in the *Ikhtīyārāt.* The models differ only in how they treat the question of the planetary latitudes. As far as motion in longitude they are the same.[93]

For this model, as far as the motion of the encompasser is concerned, Shīrāzī has the following:

> And the motion of encompasser is equal to the motion of its center, meaning the motion of the deferent of the planet such that in the upper half it is in the direction opposite that of the deferent.[94]

Thus, it is apparent that in the *Nihāya* Shīrāzī had already rejected the use of 'Urḍī's Lemma (i.e., his Deductive-Superior) in favor of a scheme based on Apollonius's Theorem (i.e., Conjectural-Superior), just as he did in the slightly later *Ikhtīyārāt.*

One of the interesting features of the two *Nihāya* manuscripts used for this study, Köprülü 956 and 957 (completed, as we saw, approximately 1 and 3 years, respectively, after the book was first written) is the evidence for extensive revisions in the discussion of the model for the superior planets appearing in the chapter on the planetary latitudes. This is particularly true of a long section in Köprülü 957 that has been crossed out entirely and supplanted by a revised text. Of interest to our discussion is the fact that in each of the two manuscripts the short statements pertaining to the orientation of the orbs have been crossed out, and a revised set of orientations appended to the marginal text. As we have said, what has been preserved in the *Ikhtīyārāt* is a reflection of the earlier version of Shīrāzī's proposed alignment of the spheres (see Appendix E).

The numerous revisions and emendations in the discussion of the planetary models (embedded as we saw as they were in the chapter on the planetary latitudes)

---

[92]Shīrāzī, *Nihāyat al-idrāk fī dirāyat al-aflāk*, Köprülü MS 956, 77v.

[93]What is different in this section of the *Nihāya* relative to the *Ikhtīyārāt*, is the orientation of the orbs. This is likely due to Shīrāzī's treatment of the involved problem of planetary latitudes. In the model for the superior planets as it appears in the *Ikhtīyārāt* the deferent and the dirigent were assumed to oriented such as to share the plane of their equators, turning on parallel axes; while the equators for the encompasser, maintainer, and the epicycle shared the same plane. In the *Nihāya* the encompasser is aligned with the deferent whereas the equator for the dirigent, maintainer, and epicycle all lie in the same plane.

[94]Shīrāzī, *Nihāyat al-idrāk fī dirāyat al-aflāk*, Köprülü MS 956, 114v.

are unexpected and indicate that the contents of the *Nihāya* was being edited by the author long after the first version of the work was published.[95] The evidence for Shīrāzī's dissatisfaction with his planetary theory is present not only in the abundant marginal notes but in the main body of the text itself. In the chapter on the latitudes, after presenting his theory for the superior planets (a theory that he revised in the *Tuḥfa*) he writes:

> So the difficulties occurring in the motion of these three have been overcome by the three orbs which we have added, and this is the figure of the corporeal orbs all three as it is possible to imagine them in a plane and this is according to the Conjectural Hypothesis, <u>not</u> according to what we have chosen.[96]

A look at this line in Köprülü 957 indicates rather plainly that the negation particle *lā*, which is squeezed into the space between its neighboring words, is a later addition (See Fig. 4.9). Indeed, this fragment would make considerably more sense if the *lā* (i.e., the "not") were to be removed. That the Conjectural Hypothesis was Shīrāzī's chosen model at the time the Köprülü 957 was originally completed (i.e., prior to the emendations by the author) is further evidenced by the fact that the subsequent text containing the emended model, a considerable fragment more than a page long, could only be added to the margins of the folio.[97] The additional text continues: "for the [correctly] imagined method is different than this, and it is that we assume for all of the superior planets and encompasser orb ... with its motion equal to the motion of the center of the epicycle for the planet and in the upper half in the sequential direction."[98]

It appears then that at the time this revision was carried out that Shīrāzī had finally settled on the Deductive Hypothesis for the planets. But even here, Shīrāzī appears to be struggling with his choice of model, since the word *khilāf* (counter) has been crossed out from the marginal text which would originally have read "with its motion equal to the motion of the center of the epicycle for the planet and in the upper half in the counter-sequential direction." It is only after many rewrites that Shīrāzī settles on the Deductive Hypothesis (i.e., one based on 'Urḍī's Lemma) as his proposed solution, by stating in a commentary upon the marginal commentary: "And as for the uniformity of the motion of the center of the epicycle about the equant and the alignment of its diameter, this is as was [described] in the principle of the Maintainer and the Dirigent."[99] At the end of the extensive revisions of the model for superior planets, immediately prior to moving on to a discussion of the

[95]Chapter 2.8 of the *Nihāya* (on the superior planets) is also one of the sections of the book that show similar evidence of revision.

[96]Shīrāzī, *Nihāyat al-idrāk fī dirāyat al-aflāk*, Köprülü MS 957, 98r.

[97]Shīrāzī, *Nihāyat al-idrāk fī dirāyat al-aflāk*, Köprülü MS 957, 98r.

[98]Shīrāzī, *Nihāyat al-idrāk fī dirāyat al-aflāk*, Köprülü MS 957, 98r.

[99]Shīrāzī, *Nihāyat al-idrāk fī dirāyat al-aflāk*, Köprülü MS 957, 98r.

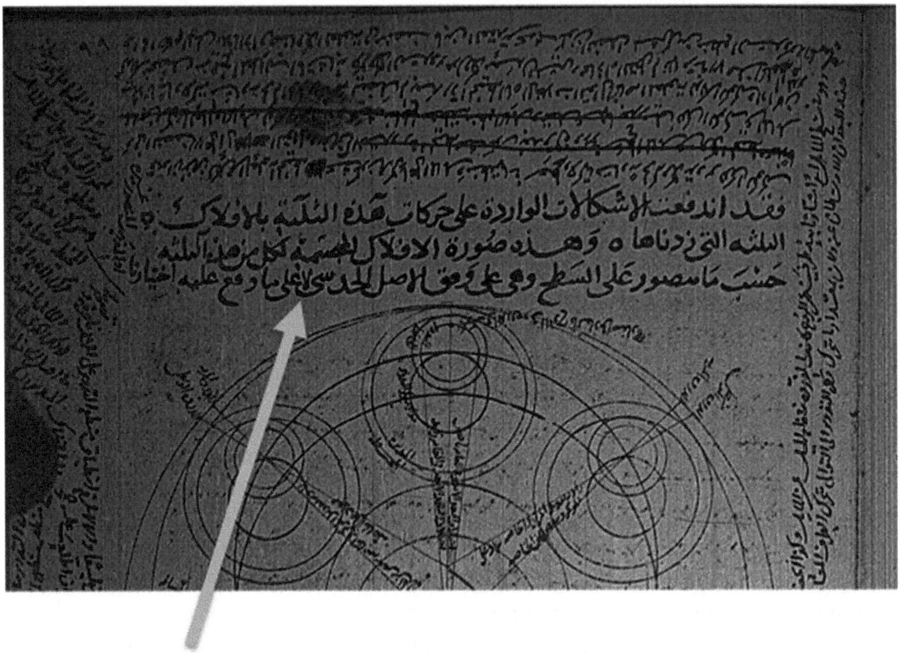

Fig. 4.9 An example of Shīrāzī's revisions in the *Nihāya* (seen here for his original model for the superior planets). These revisions include the insertion of the particle لا, negating the initial sense of his text.

inferior planets, Shīrāzī adds, tellingly: "the secret unraveled during the writing [of this tract, after] I studied it."[100]

In the chapter on the superior planets in the *Tuḥfa*, written 4 years after the *Nihāya*, Shīrāzī offers a revised model of the superior planets. This model consists of the Parecliptic, the Eccentric Deferent, the Encompasser, the Incliner (concentric with the planetary epicycle and therefore not relevant to the longitude of the epicyclic center), and finally the Epicycle.[101] As far as the motion of the encompasser he states that is "equal to the motion of the eccentric of the

[100]Shīrāzī, *Nihāyat al-idrāk fī dirāyat al-aflāk*, Köprülü MS 957, 98r. و قد انحل الرمز في اثناء التقرير ان اطلعت عليه.

[101]Shīrāzī, *al-Tuḥfa al-shāhīya*, BN Arabe MS 2516, 45v. "And so they established five orbs and five simple motions. The first orb the parecliptic .... The second the eccentric deferent in the thickness of the parecliptic such that the distance of its center from the center of the world is equal to one-half the distance between the center of the imagined deferent and the center of the world.... The third the encompasser in the thickness of the eccentric .... The fourth the incliner (mumayyila) orb enclosed within the encompasser ... with the distance of its center from the center of the encompasser equal to the distance between the centers of the eccentric and the imagined deferent for the planet as you have learned in the Third Hypothesis. Fifth the epicycle of the planet [centered] upon the center of the inclined orb."

planet in magnitude and direction in the upper half as you know from the Third Hypothesis."[102] While Shīrāzī's apparently reconsidered direction for the motion of the encompasser is consistent (at long last) with the Deductive Hypothesis, it is not at first clear what to make of his reference to the Third Hypothesis. For one thing, unlike the *Nihāya*, Shīrāzī does not number the hypotheses, i.e. the *uṣūl*, in his *Tuḥfa* (*cf.* Appendix D). A review of the chapter of the hypotheses clarifies Shīrāzī's confusing terminology, however:

> And know that of the principles requiring the third inequality, and that is the uniformity of motion of a point together with its drawing near and moving away from it is that the moving body, and let this be an epicycle, is enclosed by another which we call the encompasser in the thickness of the eccentric and with its motion equal to the motion of the deferent in magnitude and direction in the upper half.[103]

What Shīrāzī refers to as the "Third Hypothesis," in the chapter on the superior planets is apparently the hypothesis associated with the third inequality, i.e., the Deductive Hypothesis (or 'Urḍī's Lemma; see Appendix D). Given Shīrāzī's choosing of Apollonius's Theorem in the *Nihāya* and the *Ikhtīyārāt*, the manner in which he chooses to emphasize the "correct" motion for the encompasser in the *Tuḥfa* is telling:

> For if they were in different directions, and they are not, there would be drawn a circle, from the motion the center of the epicycle, through a motion compounded from the motion of the encompasser and the eccentric, with the distance of its [i.e., the circle's] center from the center of the deferent as the distance of the center of the epicycle from the center of the encompasser regardless of the supposition of the center of the epicycle at the beginning of their assumed motion at the apogee of the encompasser or the perigee, except that in the first scheme the circle is described such that its center falls [at a point] higher than the center of the deferent if the center of the encompasser is at the apogee and lower than [the center of the deferent] if the center of the encompasser is at the perigee and in the second scheme the reverse is true.[104]

This paragraph is merely a description of what Shīrāzī has previously called the Conjectural-Superior and Conjectural-Inferior hypotheses (see Figs. 4.5 and 4.6). He continues:

> If a circle is described, the desired [thing] – which is the drawing near and moving away from the point about which the motion is uniform – is not obtained... [Whereas] if the motions [i.e., of the encompasser and the deferent] are in agreement in the upper half a circle is not described, rather [this compels] the uniformity of the motion of the center of the epicycle, compounded of two motions, about a point that is separated from the center of the deferent also by the separation of the center of the epicycle from the center of the encompasser, however [this occurs] together with the drawing near to and the moving away from [the equant] as desired, regardless of the assumption of the initial location of the center

[102]Shīrāzī, *al-Tuḥfa al-shāhīya*, BN Arabe MS 2516, 46r.

[103]Shīrāzī, *al-Tuḥfa al-shāhīya*, BN Arabe MS 2516, 46r.; Morrison, "Qutb al-Din al-Shirazi's Hypotheses for Celestial Motions," 50.

[104]Shīrāzī, *al-Tuḥfa al-shāhīya*, BN Arabe MS 2516, 25v; Morrison, "Qutb al-Din al-Shirazi's Hypotheses for Celestial Motions," 50.

> of the epicycle in the apogee of the encompasser or in the perigee, the difference being that in one [scheme] the motion is uniform relative to a point above the center of the deferent and in the other [it is uniform relative to] a point below it.[105]

This, of course, is a description of what in his earlier work, the *Ikhtīyārāt*, Shīrāzī referred to as the Deductive-Inferior and Deductive-Superior hypotheses.

What we have seen of Shīrāzī's *hay'a* works indicates that his views on planetary models was in a state of flux in the period framed by the completion of the *Nihāya* and that of the *Tuḥfa*. What lends this fact its special interest, however, is that the *Nihāya* has generally been considered Shīrāzī's principal work. By the same token, the *Ikhtīyārāt* and the *Tuḥfa* have been generally viewed as derivative works that essentially restated the material in the *Nihāya*. The fact that the Shīrāzī's views were evolving, highlights his three works on astronomy as providing a rare opportunity to observe a practitioner of *hay'a* in the process of proposing, rejecting, and revising his astronomical theories.

That Shīrāzī himself was aware of his difficulties on the use of Apollonius's Theorem for his models of the superior planets is also seen in a section from the chapter on the hypotheses in the *Tuḥfa*. Here, Shīrāzī appears to reject a configuration based on Apollonius's Theorem by questioning its agreement with observation, stating: "We say [this hypothesis] was too majestic to be hidden from [Ptolemy] ... however he did not use this hypothesis because it entails matters that reality proves false."[106] After providing the reader with a list of observational inconsistencies Shīrāzī writes: "Knowing how this hypothesis necessitates these matters we have used [the hypothesis] in our books without referring to [these observations] as a test of the intellects of the intelligent [readers]: Do they pay attention to it or to some of it? And upon God is the straightness of the path and at Him the road ends."[107] Given the evidence presented in this chapter, the rather striking trail of revisions, omissions, and emendations, it is difficult not to read in these lines a concession on Shīrāzī's part to his earlier difficulties in providing an original and working model for the superior planets.

4.8 Discussion

In this chapter we have reviewed issues of chronology, and some of the structural similarities of Shīrāzī's books on *hay'a*, while examining aspects of Shīrāzī's models themselves. Even though these works still await a full scholarly study of their contents, it has – in the brief study presented in this chapter – been possible to discover a number of unexpected and surprising features. The most important of

---

[105]Shīrāzī, *al-Tuḥfa al-shāhīya*, BN Arabe MS 2516, 25v.; Morrison, "Qutb al-Din al-Shirazi's Hypotheses for Celestial Motions," 50.

[106]Morrison, "Qutb al-Din al-Shirazi's Hypotheses for Celestial Motions," 144–145.

[107]Morrison, "Qutb al-Din al-Shirazi's Hypotheses for Celestial Motions," 144–145.

these findings reflect on Shīrāzī's model-building activity. It is clear, based on what we have seen, that the period 1281–1285 C.E. was a period of intense activity in terms of Shīrāzī and his astronomical theories. Though by the time he commenced in writing the *Nihāya* he had already devised his lunar model, Shīrāzī subjected his other planetary models (most notably that of the superior planets) to heavy revisions, as is especially evident in several of the earliest surviving manuscript copies of this work, namely, MS Köprülü 957, and MS Köprülü 956 (each of which could probably be considered a new edition of the work).

From what we have seen, it is clear that each of Shīrāzī's three works on *hay'a* that were completed from 1281 to 1285 C.E. provided a possibility for the author to present his developing astronomical theories as these were being reworked and revised. The fascinating historical record of the revisions surveyed in this chapter disproves several previously published results in regard to Shīrāzī and his astronomical works. The first has to do with the date of the *Ikhtīyārāt*, and, more importantly, the significance of this book as a scientific work. Saliba dates this work to "around 1304."[108] The existence of MS Milli Library 31402 pins the completion date of the *Ikhtīyārāt* to February 1282 C.E., (i.e., 4 months after the *Nihāya*). That the *Ikhtīyārāt* is mentioned by name in the *Nihāya*, in a manuscript tradition which dates to at most 9 months after the completion of work itself, suggests as well that, in addition to their temporal proximity, the *Ikhtīyārāt* and *Nihāya* were thought of as companion works by their author.

As far as the significance of the *Ikhtīyārāt* is concerned, this work has been described both as an "abridgment of the *Nihāya*" as well as a "Persian version of the *Tuḥfa*."[109] We will look at the significance of the *Ikhtīyārāt* in some detail in the next chapter. Here, it is important to point out that the fact that the author refers readers of his principal work, the *Nihāya*, to the *Ikhtīyārāt* – in order to advise them of a more nuanced exposition of a technical topic, suggests that he considered the *Ikhtīyārāt* as an important work in its own right.

In addition, the work reviewed in this chapter, has shed new light on the situation with Shīrāzī's models and their evolution in time. The first finding here involves Shīrāzī's lunar model as he lists this in the *Nihāya*. Previous scholarship had dated the first appearance of this model to the *Tuḥfa*.[110] We have seen that this model (or variations thereof) had already been proposed by Shīrāzī in both the *Nihāya* and the *Ikhtīyārāt*. Shīrāzī statements in the *Ikhtīyārāt* chiding al-'Urḍī for not realizing the full power of his invention confirm this. Furthermore, as far as Shīrāzī's models for the superior planets were concerned, his choice of an unsuccessful model in the *Ikhtīyārāt*[111] was preceded by a similar (and similarly unsuccessful) model in the *Nihāya*.

---

[108] Saliba, "Persian scientists in the Islamic world," 141.

[109] Saliba, "Persian scientists in the Islamic world," 141 and 139.

[110] Saliba, "Arabic Planetary Theories after the 11th Century AD," 98–100.

[111] See note 79.

Finally, a side by side reading of the *Nihāya* and the *Ikhtīyārāt* has allowed for the deciphering of Shīrāzī's obscure language in regard to his "Conjectural" and "Deductive" hypotheses. With their appealing four-fold symmetry, these hypotheses appear to have lulled Shīrāzī into selecting a model (i.e., the Conjectural) for his superior planets that (as he tells us in the *Tuḥfa*) "entailed matters that reality proves false." Along with abandoning his "Conjectural" model for the superior planets Shīrāzī appears to have abandoned the nomenclature of the "Conjectural" and "Deductive" hypotheses in the *Tuḥfa.*

Chapter 5
Persian vs. Arabic: Language as Determinant of Content in Shīrāzī's Works on *Hay'a*

5.1 Introduction

This chapter examines the *Ikhtīyārāt*, and Shīrāzī's decision to render this work in his native language, Persian. This decision was contrary to the common practice of Shīrāzī's time; as has been noted, the vast preponderance of scholarly works written by scholars and scientists working in the medieval Islamic world was written in Arabic. The primacy of the Arabic language as the language of scholarly discourse held generally true regardless of the cultural background of the scientists themselves,[1] and it was an enduring phenomenon that lasted for more than a millennium. Yet, despite the dominance of Arabic as the official language of the Islamic world, there appeared within several centuries of the founding of Islam other "classical" languages within the Islamic domains; most notably Persian and Turkish.[2] In regard to the increasing importance of the Persian language, and its status as a "new dominant literary language" in what he calls the Early Middle Period (i.e., c. 1111–1274 C.E.) Hodgson states:

> [The cultural ascendance of the Persian language] served to carry a new overall cultural orientation within Islamdom. Henceforth while Arabic held its own as the primary language

[1]The illustrious Bīrūnī (973 – c. 1048 C.E.) who was born in Khwārazm, states his preference for Arabic in his book on pharmacy and materia medica, *Kitāb al-ṣaydanah fī al-Ṭibb*, by describing what was his first-hand experience of writing a scientific treatise in Khwārazmian with inadvertently humorous results: The ill-fated work appears to have elicited astonishment as that of "a camel at the rain-gutter or a giraffe at the stream."

وهى مطبوعة على لغة لو خُلّد بها علمٌ لاستغرب استغراب البعير على الميزاب و الزرافة في الكراب.

Muḥammad ibn Aḥmad Bīrūnī, *Āl-Biruniś Book on Pharmacy and Materia Medica* (Karachi: Hamdard Academy, 1973), 12; D. Boilot, "al-Bīrūnī (Bērūnī) Abu'l-Rayḥān Muḥammad b. Aḥmad," *Encyclopaedia of Islam, Second Edition* (Brill Online, 2011), http://www.brillonline.nl/subscriber/entry?entry=islam_SIM-1438.

[2]Marshall G. S Hodgson, *The Venture of Islam*, vol. 2, 293.

K. Niazi, *Quṭb al-Dīn Shīrāzī and the Configuration of the Heavens: A Comparison of Texts and Models*, Archimedes 35, DOI 10.1007/978-94-007-6999-1_5,

> of the religious disciplines and even, largely, of natural sciences and philosophy, Persian became, in an increasingly large part of Islamdom, the language of polite culture; it even invaded the realm of scholarship with increasing effect.[3]

As far as scientific productivity within the Persianate domains of Islam, E. S. Kennedy has described a trend by which, beginning in the tenth century, the cluster of productions sites for astronomical tables, or *azyāj* (sing. *zīj*), can be seen to drift eastward from Baghdād. These loci of astronomical research appear to have remained centered on the Iranian plateau, for four centuries starting from roughly 1100 C.E.[4] While noting that astronomical tables represent only one genre of scientific writing, Kennedy considers them as useful indicators of the intensity of scientific activity (by virtue of their including, simultaneously, elements of observation, theory, astronomy, and mathematics), and notes that during the twelfth to sixteenth centuries Persia was able to export scientific knowledge to its neighbors, by virtue of its dominance in the various fields of science.[5]

Reflective of the role of the Arabic language as the lingua franca *par excellence* of academic discourse in the Islamic world, it should be noted that *hay'a* texts in the Persian language are not nearly as numerous as those written in Arabic. Indeed, many of the *hay'a* works that Storey lists in his survey are described as translations of Arabic works.[6] As with other genres of scholarly writing, the reasons for the relative scarcity of Persian texts on *hay'a* presumably had to do with the authority of Arabic as the language of the *Qur'an* and the hadith, as well as the fact that by writing in Arabic authors could be assured of finding readers anywhere in the vast realms of the Islamic world.[7]

---

[3]Hodgson, *The Venture of Islam*, vol. 2, 293.

[4]Kennedy, "The Exact Sciences in Iran under the Saljuqs and Mongols," 678.

[5]Kennedy, "The Exact Sciences in Iran under the Saljuqs and Mongols," 678. Writing on the life sciences, alchemy, and medicine, S. H. Nasr places the peak of scientific activity in Persia at an earlier date: "The Islamic conquest of Persia enabled the Persians to become members of a truly international society and to participate in a worldwide civilization in whose creation they themselves played a basic role. A homogeneous civilization which spread from the heart of Asia to Europe, possessing a common religion and a common religious and also scientific language, facilitated the exchange of ideas and prepared the ground for one of the golden ages in the history of science, in which the Persians had a major share. Islamic science came into being in the 2nd/8th century as a result of the vast effort of translation which made the scientific and philosophical traditions of antiquity available in Arabic. This early phase of activity reached its peak in the 4th/10th and 5th/11th centuries just before the Saljuq domination. During this period, which is among the most outstanding in the history of science, Persia was the main theatre of scientific activity, and although there were certainly many Arab and other non-Persian scholars and scientists, most of the figures who contributed to the remarkable philosophical and scientific activity of the age were Persians." S. H. Nasr, "Life Sciences, Alchemy and Medicine." In *The Cambridge History of Iran*, Richard Frye Ed. (Cambridge, Cambridge University Press, 1999), vol. 4, 396.

[6]Storey, *Persian Literature, A Bio-Bibliographical Survey* (London: Luzac & Co., LTD., 1958) vol. 2, 1:35–117. These works, as a rule, await scholarly studies.

[7]It should be noted here, that a potential problem of writing *hay'a* works in Persian, the availability of technical terminology in Persian, was likely not an issue (at least by the fifth century A.H.), as

The increased prominence – through the Islamic era – of the Persian language, and the importance of the Iranian plateau as a locus of scientific activity, raises a number of questions in regard to Shīrāzī's decision to write the *Ikhtīyārāt* in Persian. Was this decision perhaps a reflection, of the new "cultural orientation" posited by Hodgson? Did the *Ikhtīyārāt* signal a trend or was it an isolated instance in the authoring of scientific texts in Persian? How does this work relate to *al-Risāla al-Muʿīnīya* (or the *Muʿīnīya Epistle*) and *Ḥall-i Mushkilāt-i Muʿīnīya*, two Persian texts on *hay'a* that were authored by Shīrāzī's illustrious teacher, Ṭūsī?[8] Given the existence of these earlier texts (completed nearly 50 years before *Ikhtīyārāt*), and the results presented in Chap. 4, what can be said about the *Ikhtīyārāt* and its significance as a *hay'a* text in Persian?

In looking at previous scholarship on the topic we note that Saliba has touched on all three of the Persian *hay'a* texts referenced above (i.e., *al-Risāla al-Muʿīnīya*, *Ḥall-i Mushkilāt-i Muʿīnīya*, and the *Ikhtīyārāt*) in his essay "Persian Scientists in the Islamic World."[9] Saliba begins his essay by noting the difficulty of identifying Persian scientists of the medieval world when the term Persian is viewed in an ethnic sense.[10] Rejecting, a geographical interpretation of Persia as well,[11] he opts, instead, to use the term Persian in a linguistic sense. Saliba's study is therefore, oriented toward teasing out the significance of *hay'a* works that were written in Persian against the vast number of *hay'a* texts that were written in Arabic.[12] In the ensuing discussion, Saliba sets forth his hypothesis on how the relative scarcity in the number of these Persian texts appears to have been reflected in turn by the mediocrity of their technical contents. In Saliba's final evaluation, the *Ikhtīyārāt* and the other examples of the *hay'a* genre written in Persian (such as *al-Risāla al-Muʿīnīya*, and *Ḥall-i Mushkilāt-i Muʿīnīya*) were either popularized (or in some other way debased) translations of Arabic works or ones that were "superseded" by other of their authors' works that were written in Arabic.[13]

The impression that the reader can not avoid on reading "Persian Scientists in the Islamic World," is of the sterility and impoverishment of Persian *hay'a* texts and (by extension – when taking Persian in a linguistic sense) of a negligible "Persian" imprint on a rather important and productive branch of medieval science, *hay'a*. To understand how this conclusion could be so different from that of earlier historians

---

demonstrated by Bīrūnī's work on astronomy *Kitāb al-tafhīm li-awā'il fī ṣināʿat al-tanjīm*, which was completed in both Persian and Arabic versions in 1029 C.E.

[8] As with the vast majority of scientific texts from the Islamic world, neither of these works have been edited or translated at present, and thus each awaits a full and in-depth study.

[9] Saliba, "Persian Scientists in the Islamic World." in *The Persian Presence in the Islamic World*, 126–146. Cambridge: Cambridge University Press, 1991. It should be noted here that this essay serves primarily as a survey rather than an in-depth study of the *hay'a* texts it lists.

[10] Saliba, "Persian scientists in the Islamic world," 138.

[11] As was adopted, for instance, by the aforementioned study by Kennedy.

[12] Saliba, "Persian scientists in the Islamic World," 127.

[13] The key claim is that the Persian *hay'a* texts are secondary works as far as their technical merit and significance. Saliba, "Persian scientists in the Islamic World," 138.

of the twentieth century such as E. S. Kennedy we note that Saliba's conclusion hinges on a question of semantics and the meaning assigned to the word "Persian."

It is worth examining this claim further, however. That a semantic choice would help uncover a lackluster tradition of *hay'a* writing in the Persian language is not terribly surprising at first glance. Why should it come as a surprise, in other words, that medieval Persian-speaking scientists should render their most important scientific contributions in anything but Arabic which was the lingua franca of Islamic scholarship? By the same token, if Persian was the vernacular language of the scientists in question, it (and not Arabic) would have been the obvious vehicle for the popularization of complicated astronomical concepts. Once one accepts the proffered definition of "Persian," therefore, Saliba's assertion sounds plausible enough. I hope to establish in the remainder of this chapter, however, that blanket claims as to the mediocrity of *hay'a* texts that were written in Persian are utterly unwarranted.

Before we return to the text of the *Ikhtīyārāt* to examine it with a view to its originality and significance, we leave these introductory remarks with a quote from Ibn Khaldūn's *al-Muqqadima,* from a chapter that he devotes to the *'ajam* scientists and their role in the Islamic world. Ibn Khaldūn is interested in explaining the outsized contributions of the *'ajam* scientists, and he does so partly, by claiming that at the outset the Arab conquerors of Persia were focused on matters political, i.e., on the maintenance of power and its apparatus, and thus relegated scholarly activities to the *'ajam.*[14] Ibn Khaldūn concludes his chapter in a rather striking passage that bears directly on the effect of the Mongols on the scholarly productivity of the Persian-speaking lands, and thus on our discussion of Shīrāzī and his colleagues.

> [The near-exclusive involvement of the *'ajam* with the religious and intellectual sciences] remained the case in the Islamic lands so long as civilization was in Persia and its regions in Iraq and Khurāsān and Transoxiana. And when these regions were destroyed and civilization, which is the divine secret for the obtaining of knowledge and crafts, left them, knowledge left all of the *'ajam* for they were surrounded by nomadism, and knowledge is specific to lands that are abundant in civilization/settled-living. And today no land has more abundant civilization than Egypt. So, she is the mother of the world, the īwān of Islam and the well-spring of the sciences and crafts. There remains some civilization in Transoxiana due to [the government that is there], and it can not be denied that through it they [i.e., the inhabitants of Transoxiana] have a bit of the sciences and crafts. And what has led us to this belief are one of their scholar's written works that have reached us from those lands, and this scholar is Sa'd al-Dīn al-Taftazānī [1332–1390 C.E.]. As for the other *'ajam* we have not seen after the Imam Ibn al-Khaṭīb [i.e., Fakhr al-Dīn Rāzī, (1149–1209 C.E.)] and Naṣīr al-Dīn al-Ṭūsī any works that would indicate excellence.[15]

Ibn Khaldūn's observations stand in contrast to what we have seen earlier in this chapter as far as the scientific productivity of the Perso-Islamic world in the aftermath of the Mongol invasion. How is one to reconcile these contradictory views, one of cultural annihilation and the other of immense cultural productivity – one held by a preeminent historian living in the century following the Ilkhan era, and the other

---

[14]Ibn Khaldūn, *The Muqaddima* (Cairo: Matba'at Mustafa Muhammad, 1945), 543.

[15]Ibn Khaldūn, *The Muqaddima*, 545.

by modern historians of the Middle East? Ibn Khaldūn's observations in regard to the virtual disappearance of scholarship from Persianate lands may certainly have been driven in part by a diminishment in the transmission of information from Persia to the Levant – due to the mortal enmity of the Ilkhans and the Mamluks – and was perhaps due to other factors, as well (such as a personal apathy of Ibn Khaldūn to the legacy of the Ilkhans). Certainly, that a wholesale destruction, in the early decades of the thirteenth century, of the cities of Transoxiana, Khurāsān, and other regions in Persia, would have affected the intellectual productivity of Persianate lands sounds plausible enough. That the historical evidence, to the limited extent that this has been studied, does not reflect this (reflecting instead a period of scientific productivity), is a problem for future historical scholarship to study and explain. As we recognize the contributions of Shīrāzī and his colleagues in the Persianate world to the sciences and to Islamic civilization as a whole, we should, however, appreciate the severity of the blows suffered by what Ibn Khaldūn refers to as civilization (or sedentary life) in the Persian-speaking land from which Shīrāzī hailed. Given the accounts that we saw in regard to the violence of the Mongol campaigns and the onerous economic conditions for much of the Ilkhanid era, it is possible to view the cultural achievements of Shīrāzī and his colleagues as a testament, as well, to the resilience and indefatigability of Persia, its people, and its culture.

5.2 Persian vs. Arabic in the Chapter on the Superior Planets

We turn now to the *Ikhtīyārāt* to review its significance as a work of *hay'a*. Saliba describes the *Ikhtīyārāt* as two things: an "abridgment of the author's Arabic *Nihāya*,"[16] and later in the same article as the "Persian version of the *Tuḥfa*."[17] Leaving aside the question of how one (putatively) derivative work could be extracted from two different sources, we revisit first the evidence we saw in regard to its dating. As we saw Saliba posits the *Ikhtīyārāt* as having been written "around 1304."[18] In Chap. 4 we saw that this work was completed shortly after the *Nihāya*, on Feb. 19th, 1282 C.E., i.e., 22 years prior to Saliba's date. Since the *Tuḥfa* was written *after* the *Ikhtīyārāt*, it plainly can not serve as a source for the latter.[19]

Since the *Ikhtīyārāt* could not have been based on the *Tuḥfa*, and since the *Ikhtīyārāt* and *Nihāya* are more closely related (as, for example, can be seen from the close correspondence of their planetary models), we could ask, instead: Is

[16]Storey, *Persian Literature*, vol. 2, 1:64; Saliba, "Persian scientists in the Islamic world," 141.

[17]Saliba, "Persian scientists in the Islamic world," 138.

[18]Saliba, "Persian scientists in the Islamic world," 141.

[19]In addition, unlike the other two works of Shīrāzī that we examined in Chap. 4, Shīrāzī does not include in the *Tuḥfa* a preliminary list of planetary models that he considered obsolete, instead limiting himself (generally speaking) to a presentation of the models that he accepted as legitimate. The Tuḥfa is a generally more compact book than either the *Ikhtīyārāt* or the *Nihāya*, and it would be difficult to see how this work could serve as the source of a further summarized rendition of the same material.

the *Ikhtīyārāt* a translation (or perhaps a popularized version) of the *Nihāya*? In considering this question it is important to recall that we have previously discussed how the *Nihāya* predates the *Ikhtīyārāt* by 4 months, and also that Shīrāzī claims in his introduction to the *Ikhtīyārāt* that this work is an "adorning" of the text of the *Nihāya* by the Persian language.

Yet, here as well, there are difficulties with the view of the *Ikhtīyārāt* as consisting of a rendition of the *Nihāya* into Persian (Shīrāzī's claims to the contrary). How can the *Ikhtīyārāt* be a translation of the *Nihāya* into Persian, when, as we saw, the two books differ in Shīrāzī's treatment of the hypotheses (see section 4.5, and Appendix D)? Recall, that the material systematically presented in the section on the ḥypotheses in the *Nihāya*, appears scattered about the *Ikhtīyārāt*, partly ending up in the chapters on the Sun (e.g., *the Hypothesis of the* Dirigent), the Moon (e.g., the Conjectural Hypothesis, and the Ṭūsī Couple) and the superior planets (e.g., variants of the Conjectural and Deductive hypotheses). Also, as we noted in Chap. 4, Shīrāzī's labeling scheme for his hypotheses are different in the two works (note in particular the *Nihāya* hypotheses labeled 5 and 8 in Appendix D). It is important to note here, as well, that Shīrāzī chose to include the discussion of Venus's longitudinal motion in the chapter on the superior planets in the *Nihāya* (likely following Ptolemy's scheme in the *Almagest*), but, in the *Ikhtīyārāt,* this material appears alongside the longitudinal models for Mercury.

An even more notable difference in the chapters on the superior planets as they appear in the two works, however, is the fact that in the *Nihāya* Shīrāzī proposed his planetary models *after* presenting his discussion on the planetary latitudes, whereas for the *Ikhtīyārāt* these appear each in their properly designated chapter. As a result of these changes the outline for the chapter on the superior planets are notably different (see Appendix B, section 2, 6 and 7).

As we have said Shīrāzī's original models, though not identical (and though appearing in different parts of their respective books), correspond closely to each other in these two works (see Chap. 4, Sects. 4.6 and 4.7).[20] If we were to ignore the differences listed in the previous paragraph, and look *only* at the differences in the models for the superior planets, we could begin to speculate about a relationship between these two works other than one based on a putative "translation from Arabic to Persian." If, rather than taking the author's Shīrāzī's words as to the nature of the *Nihāya* as a book containing his mature and fully-developed thinking relative to his models, we look at it as a work in progress, containing his views on *hay'a* at the beginning of a period of intellectual ferment and productivity, then the fact that he wrote two books in close succession (three, counting the *Tuḥfa* of 1285 C.E.) could more accurately be described as a process by which the author consigned to paper his evolving theories with respect to astronomy at three closely separated instances in time. In this view which is lent credence by the heavy revisions that appear in Köprülü 956 and Köprülü 957, as well as the grossly different model for

[20]We have already commented in Chap. 4 on how the differences in the two books are primarily in the relative orientation of the various axes of rotation for the orbs of the superior planets.

the superior planets that was to subsequently appear in the *Tuḥfa*, each of the three books in question would serve as a "snapshot" of Shīrāzī's thinking relative to his astronomical models over a 4 year period.

To provide a rationale for the choice of language in the *Ikhtīyārāt*, we would, of course, have to speculate. Could this have been driven, perhaps, by practical considerations such as the desire to locate a patron who was conversant in (or at least familiar with) the language of the text? As is self-evident, Shīrāzī was capable of writing in both Persian and Arabic, and he may very well have taken the opportunity afforded by a sponsor who did not understand Arabic to compose his work instead in Persian (or perhaps to organize and publish notes that he had already produced).

So far in our discussion we have focused primarily on the difference in the organization of the material within the two books, i.e., on material that ended up at a different chapter or location for each. A look at Appendix B suggests that there are more substantial differences in the two texts. Indeed, each work contains extended sections that are omitted in the other.[21] Of several notable examples of sections that appear in the *Ikhtīyārāt* or the *Nihāya* but not the other we will now look at three in some detail.

5.2.1 The Eccentricity of the Equant and of the Deferent

The *Nihāya* and the *Ikhtīyārāt* both contain an extended section in which Shīrāzī discusses his ill-fated Conjectural Hypothesis (recall that Shīrāzī abandons the Conjectural Hypothesis subsequently, and instead provides a list of its observational inconsistencies in the *Tuḥfa*). The context for these arguments in favor of the Conjectural Hypothesis (see Appendix D, item 8 in the columns corresponding to the *Nihāya* and the *Ikhtīyārāt*) is Ptolemy's discussion of the eccentricities of the equant and the deferent orbs. In the *Almagest*, Ptolemy posits the eccentricity of the deferent (i.e., the distance of the center of the deferent orb with respect to the center of the World) to be one-half the eccentricity of the equant (i.e., the separation of the equant point form the center of the World). Ptolemy is not clear as to the reason for his claim, and scholars have speculated as to the reasoning behind his assertion ever since.[22] Rather than provide an explicit reasoning, Ptolemy states cryptically

[21] An example of this is a fragment that provides an alternative explanation for the necessity of existence of an epicycle, and appears in the *Nihāya* but not in the *Ikhtīyārāt*. "And as for the [possibility] of retrograde motion and all that it entails, without the presence of an epicycle, though [referred to previously] in the Fourth Hypothesis, we will [nonetheless describe it in a different manner, which will include benefits that the aforementioned [discussion] lacked." Shīrāzī, *Nihāyat al-idrāk fī dirāyat al-aflāk*, Köprülü MS 956, 54r. As this fragment is a short elaboration of a point that Shīrāzī had already made, it is not terribly interesting. Examples of more substantial variations in these two works are presented in the three sections following the current one.

[22] Pedersen, *A Survey of the Almagest*, 266–267; James Evans, *The History & Practice of Ancient Astronomy* (New York: Oxford University Press, 1998), 357.

that for the motions of Mars, Jupiter, and Saturn, and "using rough estimation, the eccentricity one finds from the greatest equation of ecliptic anomaly turns out to be about twice that derived from the size of the retrograde arcs at greatest and least distances of the epicycle."[23] Evans suggests that Ptolemy must therefore have calculated the eccentricity needed to provide a reasonable prediction of the motion of the epicyclic center about the center of the world (the so-called zodiacal anomaly, or, as Ptolemy states "greatest equation of ecliptic anomaly") as well as calculating an eccentricity required to properly predict the synodic behavior of the planet as characterized by the "size of the retrograde arcs."[24] According to Evans, the fact that these two quantities were related by a factor roughly of two is likely what led Ptolemy to specify the 1:2 ratio as an exact ratio for his models of Venus, Mars, Jupiter, and Saturn.[25]

In defending his configuration which like 'Urḍī's configuration of the superior planets relies on an eccentric deferent that is centered not where Ptolemy has placed it, but half-way between Ptolemy's center (for the deferent) and the equant point, Shīrāzī includes a passage in the *Ikhtīyārāt* that appears to be a paraphrase of the view of other astronomers who suspected Ptolemy's choice for the eccentricity of the deferent to be fixed (by calculation) and who reproached Shīrāzī for moving the center of the deferent from were Ptolemy had placed it. Mirroring what was perhaps the formal disputational scheme of his day Shīrāzī paraphrases the criticism as follows:

> Even though it is generally accepted that Ptolemy determined the distance of the center of the deferent for these planets by guessing ... unlike his derivation for the location of the equant [which is based on proof], this is ... false, for his reasoning there was also based on [geometrical] proof and observation. However since the proof was not listed in the *Almagest* people assumed falsely that he had determined the aforementioned distance by conjecture and by guessing, whereas this is not the case. And just as one shouldn't alter the distance between the equant from the center of the world [from that which Ptolemy has determined], one should also not change the distance of the center of the deferent [from the center of the world] for the basis of both is [a geometrical] proof.[26]

Shīrāzī's response appears immediately as follows:

> We reply that the proof indicates that the distance between the mid-point between the furthest and the closest distance of the center of the epicycle from the center of the world, that was determined from the largest and smallest arcs of the retrograde in the ecliptic, was half that between the center of the world and the equant, and we have not moved this point from its place, but we changed the distance of the center of the embodied deferent from that which the moderns [!] had set, and there is no problem with that since their basis in this [choice] was not observation or proof, nor was it the basis of Ptolemy in [assuming that the center of the epicycle was always moving along a circle centered at the point that was the bisector of the farthest or nearest distance].[27]

[23] Ptolemy, *The Almagest*, 480.

[24] Evans, *The History & Practice of Ancient Astronomy*, 358.

[25] Evans, *The History & Practice of Ancient Astronomy*, 358.

[26] Shīrāzī, *Ikhtīyārāt-i Muẓaffarī*, Ayasofya MS 2575, 116r.

[27] Shīrāzī, *Ikhtīyārāt-i Muẓaffarī*, Ayasofya MS 2575, 116r.

What is remarkable about this passage (that also appears with some variations in the *Nihāya*) is that it both highlights the existence of a current within the Islamic tradition that explained Ptolemy's choice for the eccentricity of the deferent sphere based on a measurement of retrograde arcs (as Ptolemy himself had hinted), as well as providing an insight into Shīrāzī's critique of Ptolemy. If Ptolemy's derivation of the eccentricity of the deferent was based on the observation of the arcs of retrograde, then this did not automatically support his claim of a deferent sphere with a prescribed eccentricity (which, to Shīrāzī, remained unsupported). In other words, if one were to posit another configuration (as had al-'Urḍī) that could predict the same behavior as far as the length of the arcs of retrograde were concerned, then this configuration was as valid as Ptolemy's, and eccentricities different from Ptolemy's could therefore be allowed.

As we have said the quoted text of the *Ikhtīyārāt* follows, with variations, that of the *Nihāya.* The *Nihāya* develops this idea further, however, positing a procedure for Ptolemy's derivation of the eccentricity of the deferent orb from observations of the retrograde arcs of the planet:

> So we say, and [to God we look for success], that Ptolemy obtained through observations of successive years the amount of retrograde, meaning the degrees by which the planets retrograded from first station to second station until he found from the amount of the retrogrades the smallest and the largest and he inferred from the smallest that the center of the epicycle was at the apogee [i.e., of the deferent] at the midpoint of the retrograde and from the largest that it was at the perigee [likewise], relying on the fact that – upon the limiting of the distance from both directions – should there [exist the least bit of discrepancy] that there should not befall the calculation a noticeable error due to this. He then started from the knowledge of these two quantities to seek the desired quantity in the manner which I will follow.[28]

The "desired quantity" referenced in the quote is the eccentricity of the deferent orb mentioned above. Shīrāzī subsequently proceeds with a mathematical derivation in which he extracts the eccentricity of the equant, and that of the deferent based on the measure of the largest and smallest arcs of retrograde, and a single observation of the planet, in opposition, at ninety degrees from the apsidal line.[29] Shīrāzī's derivation is related to the material in the *Almagest* X6, and X7, in which, Ptolemy derives the eccentricity of the equant and its location relative to the equinox for Mars based on observations of the planet at three solar oppositions. This section of the *Nihāya* has been studied by Gamini and Masoumi Hamedani in a recently published article.[30] Of special relevance to our discussion in regard to the differences between the *Nihāya* and the *Ikhtīyārāt* is that this derivation and its accompanying figure are missing

[28] Shīrāzī, *Nihāyat al-idrāk fī dirāyat al-aflāk*, Köprülü MS 956, 60v.

[29] Shīrāzī, *Nihāyat al-idrāk fī dirāyat al-aflāk*, Köprülü MS 956, 60v.

[30] Amir Mohammad Gamini and H. Masoumi Hamedani, "Al-Shīrāzī and the Empirical Origin of Ptolemy's Equant in His Model of the Superior Planets," *Arabic Sciences and Philosophy* 23, 1 (2013): 47–67. I am most grateful to Dr. Gamini and Professor Masoumi Hamedani for providing a copy of their paper prior to its publication.

from the *Ikhtīyārāt*. What we get in the *Ikhtīyārāt* instead is a short description of the omitted material:

> And know that [relying] on the measures of the retrogrades ... Ptolemy extracted the distance of aforementioned bisector [of the nearest and farthest distance of the epicyclic center from the center of the world] and assumed an imaginary circle centered upon it, and imagined that the center of the epicycle was always moving upon this circle, and then he observed the center of the epicycle in the mean ḍistance relative to motion and from the angle [between] the two apogees [i.e., the mean and the visible], which is at its maximum at that point, and which he determined by observation he extracted the distance between the center of the world, and the [eccentric], ... and it [was] twice the original quantity.[31]

The fact that the actual proof was omitted in the *Ikhtīyārāt* is consistent with the view of this work as a summary of the *Nihāya* (a view which this chapter aims to debunk), especially since what appears in the *Ikhtīyārāt* is an outline of the discussion that appears in the *Nihāya*. We should bear in mind that even in the *Nihāya* Shīrāzī's derivation is unsubstantiated; it remains a purely abstract exercise and does not refer to actual values or planetary parameters.[32]

Also worth noting in regard to Shīrāzī's derivation of the equant is that while it finds a full presentation in the *Nihāya*, it does appear in the *Ikhtīyārāt* as a paraphrased summary, at least. We will now examine two sections from the *Ikhtīyārāt* chapter on the superior planets that are wholly missing from the *Nihāya*.

5.2.2 *The Conjectural and Deductive Hypotheses*

After concluding his discussion of the various anomalies for the superior planets, in the *Nihāya* Shīrāzī provides a brief discussion of the equant before stating: "And should a problem arise, we respond that the reason for the motion of a [moving body] about a point that is not the center of its mover is one of three hypotheses."[33] He then proceeds to describe (1) The Ṭūsī couple, (2) his own ill-fated "Hypothesis of the Maintainer and the Encompasser," (which is how he refers to his own Conjectural hypothesis in the *Nihāya*) and (3) the hypothesis based on 'Urḍī's Lemma (the Deductive hypothesis).[34] Since he does not present his own "preferred" model until later in this work, Shīrāzī doesn't make any evaluative statements as to which of these hypotheses are acceptable as far as the configuration of the superior planets is concerned.

---

[31] Shīrāzī, *Ikhtīyārāt-i Muẓaffarī*, Ayasofya MS 2575, 116v.

[32] Indeed, Gamini and Masoumi Hamedani demonstrate the existence of an error in Shīrāzī's derivation. Shīrāzī does not include this derivation in the *Tuḥfa*. (See note 30).

[33] Shīrāzī, *Nihāyat al-idrāk fī dirāyat al-aflāk*, Köprülü MS 956, 58v.

[34] Shīrāzī, *Nihāyat al-idrāk fī dirāyat al-aflāk*, Köprülü MS 956, 58v.

While the presentation of the corresponding material that appears in the *Ikhtīyārāt* is generally similar, it differs in an important way. In this chapter, Shīrāzī has already committed to the "Conjectural Hypothesis" (which corresponds to what, in the *Nihāya* he calls the "Hypothesis of the Maintainer and the Encompasser,") as his preferred model. His approach for the presentation of the material is, therefore, to present the other two models and to demonstrate (or at least imply) the shortcomings of each:

> And since [we have reached this point and you are already aware] that these planets don't have an equant problem or an alignment problem, thanks to an interpretation that is uniquely ours and likewise in the refutation of the issue of the equant as was described in the Conjectural Hypothesis, it is time now to mention that which has reached us from the experts in this art as far as the refutation of the issue of the equant in these planets [i.e., Saturn, Jupiter, and Mars], so that beginner's don't consider these discussions by the experts as complete and so they don't come to believe in them as the final truth.[35]

We see clearly, then, that though the presentation of the material in the *Ikhtīyārāt* is similar to that of the *Nihāya*, it is also different in that (unlike the presentation in the *Nihāya*) the hypotheses based on Ṭūsī and al-'Urḍī are presented here as expressly flawed.

The discussion of this material as it appears in these two books is different in other important ways as well. In his presentation of 'Urḍī's Lemma in the *Nihāya* Shīrāzī states: "And the Third Hypothesis is what I promised to you I'd explain when needed, and that is the Hypothesis of the Maintainer and the Dirigent, that is one of the four hypotheses that are apparent from the words of Ptolemy."[36] Nowhere in the *Nihāya* does Shīrāzī subsequently explain his cryptic reference to the "four hypotheses" of Ptolemy. To solve the mystery of the four hypotheses, one has to refer, instead, to the *Ikhtīyārāt*. The four hypotheses are, of course, none other than the Deductive-Superior/Inferior, and Conjectural Superior/Inferior, that we saw in Chap. 4. They are presented in the *Ikhtīyārāt* at the conclusion of the presentation of 'Urḍī's Lemma in the chapter on the superior planets. We have already examined this text in Chap. 4 and so here merely present Shīrāzī's concluding remarks which highlight the importance of this four-fold scheme to the author:

> And it is obvious that these four hypotheses are as four branches belonging to Ptolemy. And though this is apparent to some, for most it will not become clear unless full consideration is given to it. This then is the heart of this matter, and from it our mediatory actions, our conjecture, and the excellence of our reasoning become apparent.[37]

Shīrāzī then proceeds with a criticism of al-'Urḍī for the failure to recognize the applicability of his own hypothesis to the case of the Moon.[38] He then provides a rather cryptic clue as to why he rejected the use of the Deductive Hypothesis for the superior planets, despite his success in applying it to the case of the Moon:

---

[35]Shīrāzī, *Ikhtīyārāt-i Muẓaffarī,* Ayasofya MS 2575, 112v.

[36]Shīrāzī, *Nihāyat al-idrāk fī dirāyat al-aflāk*, Köprülü MS 956, 59v.

[37]Shīrāzī, *Ikhtīyārāt-i Muẓaffarī,* Ayasofya MS 2575, 115r.

[38]Shīrāzī, *Ikhtīyārāt-i Muẓaffarī,* Ayasofya MS 2575, 115r.

> And the difference between these hypotheses is that the Conjectural Hypothesis results in points on the trajectory of the planet to be equidistant from the equant point ... and the center of the epicycle to have a true [i.e., circular] trajectory whereas the other two [i.e., the 'Urḍī's Lemma and the Ṭūsī Couple] do neither of these things, and for this reason the Conjectural Hypothesis is closer to the truth.[39]

The deviation of the planet's trajectory from a circular path is, then, that with which Shīrāzī faults the other two principles, and is apparently one of the reasons he decides upon using the Conjectural Hypothesis for the superior planets in his two earlier works: the *Nihāya* and the *Ikhtīyārāt*. The context in which this discussion is presented, the four-fold hypotheses of the Conjectural-Superior/Inferior and the Deductive-Superior/Inferior is almost entirely absent from Shīrāzī's the *Nihāya*, and is instead to be found rendered in Persian in his slightly later work, the *Ikhtīyārāt*.

5.2.3 The Question of "Alignment"

In concluding his discussion on the anomalies of the superior planets in the *Nihāya* Shīrāzī states:

> And the issue mentioned in the chapter on the Moon, caused by the uniformity of motion of the center of the epicycle about a point distinct from the center of its deferent is [applicable exactly] to these four planets [i.e., Saturn, Jupiter, Mars, and Venus], as well. But, as for that which was mentioned in regard to the anomaly of the alignment, that is not applicable, by virtue of the alignment [for these four planets] being relative to a point [about which] the uniformity of motion is reckoned.[40]

Here Shīrāzī is comparing the models of the superior planets with that of the Moon, with regard to two of their perceived Ptolemaic shortcomings. The first, the equant, behaves similarly, according to Shīrāzī, in both the Moon and the superior planets, whereas the second, the "alignment," is problematic.[41] Shīrāzī continues:

> And this too is apparent [though subtle] and we will clarify it further should we [encounter] it in the future, God willing. And its true reason is uniformity [of motion], since: for every sphere, the center of which is moving uniformly about a point, [there exists by necessity] a diameter that is aligned to that point, regardless of whether that point is the center of its orbit or not. And we have explained this in detail in the *Ikhtīyārāt* and you should pay heed to it, if you would like to be informed of it.[42]

We have already encountered this remarkable passage in Chap. 4, where we used the reference to the *Ikhtīyārāt* as evidence that it was written shortly after the *Nihāya*.

---

[39]Shīrāzī, *Ikhtīyārāt-i Muẓaffarī*, Ayasofya MS 2575, 115v.

[40]Shīrāzī, *Nihāyat al-idrāk fī dirāyat al-aflāk*, Köprülü MS 956, 58r.

[41]Indeed in the Ptolemaic system, the Moon and the other planets all feature "equant," that is a point about which they exhibit uniform angular motion and that is distinct from the center of their deferent orb. As far as the alignment point, i.e., the point that serves to define the mean apogee, they are different. In the planets other than the Moon, this point is defined to be identical to the equant, whereas in the Moon it is a distinct point, the so-called "prosneusis point."

[42]Shīrāzī, *Nihāyat al-idrāk fī dirāyat al-aflāk*, Köprülü MS 956, 58r.

One of the most interesting aspects of this fragment is that material that is relevant to Shīrāzī's discussion of the "alignment" does not appear in the *Nihāya* and that he, instead, refers his readers to the *Ikhtīyārāt.* The second interesting aspect is Shīrāzī's apparently general claim in regard to the alignment point, and that this be coincident with the equant for every planetary configuration, including that of the Moon.[43]

Indeed, the discussion that Shīrāzī is referencing appears in the chapter on the "hypotheses" in the *Ikhtīyārāt.* In that section Shīrāzī addresses those astronomers who like Ibn al-Haytham "expound on the corporeality of the orbs ... so that they [assign] for each motion an orb that is its mover" and declares anew the necessity of the configuration of the orbs to be such as to both agree with observation as well as being consistent with the principles of *hay'a.*[44] Shīrāzī continues: "And as for Ptolemy, who is the founder of the principles and the master of observation, since he doesn't account for bodies, and is content, rather, with presenting lines and circles according to his own goals, he is exempt from this [task]."[45] Shīrāzī then proceeds to describe the configurations of the Moon and the other planets according to traditional *hay'a* (derived ultimately, as we said, from the *Almagest*).[46] He concludes his presentation by stating: "This is the configuration of the corporeal orbs as commonly known."[47] What follows is an involved discussion in which Shīrāzī argues for why the aforementioned configurations "can not yield the desired result" while presenting what appears to be the groundwork for a novel theoretical scheme involving the coincidence of the equant and the *prosneusis* point, that he will be proposing shortly. A complete translation of these remarkable pages has not been attempted here, what is presented instead is merely an outline.

In describing the physical basis for his *hay'a* Shīrāzī states:

> It is [a given, and a thing that sound minds will also vouch for,] that [for] every circle, the circumference of which carries the center of another circle [i.e., the epicycle] and that moves with a simple and uniform [rotational] motion, moving the epicycle ..., the center of the deferent must possess three characteristics: first, the equality of angles resulting from equal motion about it; second, the equality of the distance of the center of the epicycle from [the center of the deferent in every instance]; and, third, the alignment of a specific diameter on the carried circle [with the center of the deferent circle]. This is because if the first characteristic is absent, either the circle is not a true circle or the center of the circle not a true center. And if the second characteristic is absent the motion is not uniform[!]. And if the third characteristic is absent, then a line passing through the point of intersection of the [epicycle] with one of the tangent circles and the center of the epicycle ... will not necessarily pass through the center of the other....[48]

---

[43]The coincidence of the alignment point and the equant holds true for the Ptolemaic configurations for the superior planets, but not so for the Moon.

[44]Shīrāzī, *Ikhtīyārāt-i Muẓaffarī*, Ayasofya MS 2575, 67v.

[45]Shīrāzī, *Ikhtīyārāt-i Muẓaffarī*, Ayasofya MS 2575, 68r. Shīrāzī appears to be quoting Ṭūsī directly. See *Ḥall-i mushkilāt-i mu'īnīya*, Majlis MS 6346, 219r.

[46]The models presented here correspond to the ones that he presents in the chapters for the Moon and the planets in the *Nihāya.*

[47]Shīrāzī, *Ikhtīyārāt-i Muẓaffarī*, Ayasofya MS 2575, 69r.

[48]Shīrāzī, *Ikhtīyārāt-i Muẓaffarī*, Ayasofya MS 2575, 69r.

Shīrāzī continues by noting that the commonly proposed *hay'a* of the Moon and of the other planets can not be correct since they fail to satisfy the physical requirements laid out above:

> And since [the behavior of the planets] that has been determined through observation does not result from their [proposed] configurations their configuration [can not be correct] and the effort [of the proponents of these models] is fruitless and their endeavors futile.[49]

Highlighting the key issues for these models as that of the equant and what Shīrāzī terms the "issue of alignment," he states:

> And it is [commonly accepted] by the practitioners of this craft that in these five planets [i.e., the Saturn, Jupiter, Mars, Venus, and the Moon] there exists an issue with the equant but not of alignment, and they justify this by [stating] that alignment is with a point about which motion is uniform. Yet this claim [is only complete] if they explain how it is that whenever the motion of the center of the [epicycle] is uniform about a point it is necessary that a specified diameter from the moved circle be aligned with that point. [Yet] none of the practitioners of this craft [have expressed this] or if they have it hasn't reached us, and the master of this craft [i.e., Ptolemy] merely assumed this based on conjecture since in the *Almagest* he said: "What is necessary is that the point that is the origin for the motion of the epicycle be a prescribed point which we assumed to be the apogee. The assumption that the apogee and perigee that is opposite to it are always upon a line from the center of the epicycle to the point about which the motion is uniform [is also speculative], and we found this was as we had supposed in the [superior planets and Venus] but not so in the Moon since [for the Moon] the uniformity is relative to the center of the World [but] the alignment is with the *prosneusis* point." This is the exact rendition of what Ptolemy says and what he meant by this is clear.[50]

The related material in the *Almagest* appears in section V.5, in the discussion of the "direction" of the diameter of the Moon's epicycle in which Ptolemy states: "Every epicycle must, in general, possess a single, unchanging point defining the position of return of revolution on that epicycle. We call this point the 'mean apogee,' and establish it as the beginning from which we count motion on the epicycle."[51] Ptolemy continues his discussion by noting how the mean apogee for the Moon is to be reckoned differently: "Now in all other hypotheses [i.e., all planetary configurations other than that of the Moon], we see absolutely nothing in the phenomena which would count against the following … [that] the diameter of the epicycle through the above apogee [i.e., the 'mean apogee'] … always point toward the center of revolution, at which furthermore, equal angles of uniform motion are traversed in equal times. In the case of the Moon, however, the phenomena do not allow one to suppose that."[52] Indeed, the mean apogees for the epicyclic orb of all planets, save that of the Moon, are defined as being aligned with the equant.[53] It is only in the case of the Moon in which the "mean apogee" is configured so that it

---

[49]Shīrāzī, *Ikhtīyārāt-i Muẓaffarī*, Ayasofya MS 2575, 70r.

[50]Shīrāzī, *Ikhtīyārāt-i Muẓaffarī*, Ayasofya MS 2575, 69r.

[51]Ptolemy, *Almagest*, 227.

[52]Ptolemy, *Almagest*, 227.

[53]Pedersen, *A Survey of the Almagest*, 192, 287, 303, 317.

aligns with a point that is neither at the center of the ecliptic, nor at the center of the deferent, but "a point removed from [the center of the deferent] towards the perigee of the [deferent] by an amount equal to the [eccentricity of the deferent]," i.e., the so-called *prosneusis* point.[54]

In his critique of Ptolemy's description of the *prosneusis* point Shīrāzī states:

> We say [i.e., in response to Ptolemy] that whatever is chosen as the origin of the motion of the moving body must be stationary relative to the moving body so that the distance of the moved body and its proximity be [confined] to that which is caused by its motion [i.e., the motion of the moved body and so the motions remain orderly]. So the point that is chosen as the origin for the motion of the epicycle must be stationary relative to the planets [!], and this is the import of Ptolemy's words: "It is necessary that the point be specified" meaning that it does not vary or change. However, the specified point [in this account] is nothing but the two endpoints of the diameter that is aligned with a point about which the motion is uniform, and the existence of this diameter is, in this case, necessary, since whenever the motion of the center of the epicycle is uniform about a point it is necessary that a prescribed diameter of the epicycle be always aligned with that point ...[55]

At this point Shīrāzī makes a remarkable claim:

> If the [equant] point is the center of the orbit of the epicycle as is the case with the five planets, and this will be explained in its place, this principle will be self-evident.... And [it holds also] if the point is not the center of the orbit of the epicycle, as is the case with the Moon, since the center of the world is not the center of its orbit, either because it draws near and far from it, or because [the center of the epicycle] does not have a true orbit, meaning that it does not move along a true circle, as will be explained in its proper place.[56]

Shīrāzī's reference to the equant point being the center of the orbit of the five planets appears to be a reference to those of his models relying on the Conjectural-Superior Hypothesis (see Fig. 4.5). He has yet to describe his own planetary models and so here merely promises that this will be done in the proper place. The reference to the trajectory of the Moon not being a true orbit (i.e., not being truly circular) is clearly a reference to the Deductive-Inferior hypothesis (see Fig. 4.8). As has been noted this trajectory deviates (minutely) from a circle. Still referring to the Moon, and to the same deviation of the lunar orbit from a perfect circle, Shīrāzī adds:

> And from this it becomes apparent that that which is commonly [accepted] as far as the uniformity of the distance [of the epicyclic center] from the center of the deferent is a falsehood, [for it is only necessary] that a specified point of the epicycle be at all times aligned with the center of the world.[57]

Shīrāzī then proceeds to argue for his astonishing claim (which, it should be noted, appears to be at odds with the physics of solidly rotating spheres). He concludes his

---

[54]Ptolemy, *The Almagest*, 227, Pedersen, *A Survey of the Almagest*, 189–193.

[55]Shīrāzī, *Ikhtīyārāt-i Muẓaffarī*, Ayasofya MS 2575, 70v. In describing the point chosen as the origin of motion as "stationary" Shīrāzī appears to be echoing al-'Urḍī. See al-'Urḍī, *Kitāb al-hay'a*, 110.

[56]Shīrāzī, *Ikhtīyārāt-i Muẓaffarī*, Ayasofya MS 2575, 71r.

[57]Shīrāzī, *Ikhtīyārāt-i Muẓaffarī*, Ayasofya MS 2575, 71r.

discussion by re-emphasizing his contention that "the [statement] that it is always the same specified diameter from the epicycle that is aligned with the *prosneusis* point is a well-accepted falsehood."[58]

What is left for Shīrāzī to demonstrate is the manner in which his radical re-imagining of the prosneusis point could be made consistent with Ptolemy's derivation in the *Almagest* 5.V. He does this by outlining Ptolemy's derivation, stating:

> If they say that [Ptolemy's calculation] is proof that the true apogee is fixed and the mean apogee variable, ... [and that he derived the argument of the epicycle] and found this to be 14 [degrees] and a fraction whereas according to calculation this was supposed to be 26 [degrees] and a fraction, and so he [chose his prosneusis point accordingly] and that this proves that the true apogee is variable and the mean apogee variable, we say in response that the lack of agreement between observation and theoretical prediction [in the scheme in which the true apogee is considered fixed], is not due to the variability of the true apogee, rather it is due to observational difficulties and the fact that equal arcs of the epicycle will appear different according to their distance or proximity....[59]

Shīrāzī concludes by stating:

> For this reason [i.e., observational factors] it is impossible, once the origin is [chosen as the] true apogee, for observation to match [numerical prediction], but this is not due to the [variability of the true apogee, just as] the agreement between [prediction and observation when the origin is chosen as the mean apogee] is not due to the fact that this [mean apogee] is fixed.... So based on this discussion it is apparent that the situation with the Moon is as that of the five wandering planets, by virtue of the fact they all have an "equant" problem [but none has] an alignment problem, since for each the alignment is with the point about which the motion is uniform, and the alignment with a [separate] alignment point an impossibility.[60]

The details of what Shīrāzī's novel interpretation of the "alignment" meant for his full models, as they appear in the *Nihāya* and the *Ikhtīyārāt* will have to await future studies. There is clearly much in the foregoing discussion that requires clarification, ideally in the context of a fully edited and translated version of the *Ikhtīyārāt*. What can be said about Shīrāzī's discussion, without going into the details of his "alignment" scheme, however, is that this scheme, which is based physical laws that appears to be Shīrāzī's own (among them, that alignment points be coincident with the point about which the angular motion of a heavenly body is uniform), deviates from the physics of solid spheres, by which a body remains equidistant from the center of its deferent. It is also clear that the novel physical principle that Shīrāzī proposes is closely tied both to the original models that he proposes in the *Ikhtīyārāt*, and to his conceptualization of the Deductive and Inductive hypotheses. Since Shīrāzī was forced to abandon his model for the superior planets in the *Tuḥfa*, it is likely that his proposed scheme for the alignment (as it appeared in

[58] Shīrāzī, *Ikhtīyārāt-i Muẓaffarī*, Ayasofya MS 2575, 72v.
[59] Shīrāzī, *Ikhtīyārāt-i Muẓaffarī*, Ayasofya MS 2575, 72 v.
[60] Shīrāzī, *Ikhtīyārāt-i Muẓaffarī*, Ayasofya MS 2575, 72v.

the *Ikhtīyārāt*) was similarly discarded. As far as Shīrāzī's account of the alignment point in the *Tuḥfa* itself, this does not appear to have been more successful.[61]

The details of what Shīrāzī's interpretation of the "alignment" meant for his full models will have to await future studies. For the purposes of our discussion of language and scientific production in the Islamic world, this section of the *Ikhtīyārāt* is important, however, because it is a record of Shīrāzī's thoughts on "alignment" during the period in which he was composing the *Nihāya* and the *Ikhtīyārāt*. This text appears in the *Ikhtīyārāt* and not the *Nihāya*. That Shīrāzī refers the reader of the *Nihāya* to the *Ikhtīyārāt* indicates that he considered the latter to contain his most complete exposition of the subject.

5.3 Discussion

A comparison of the *Nihāya* and the *Ikhtīyārāt* reveals a rather complex relationship between these books that defies our attempts at categorization. It should by now be clear that the *Ikhtīyārāt* can not be viewed as a translation of the *Nihāya* (i.e., from Arabic to Persian) in any recognized sense of the word; the texts (again, as seen in the chapter on the superior planets) are simply too different from each other. Nor can the *Ikhtīyārāt* be viewed as a popularization of the *Nihāya*; as we have seen from the fragments of text the Persian of the *Ikhtīyārāt* cited in this chapter matches well the tone and technical level of the Arabic of the *Nihāya*. As we have also seen each book contains technical passages (in which the author develops his ideas on the planetary configurations, or expands on them) that do not appear in the companion work. In addition we have the rather remarkable case of cross-referencing in which each work mentions the other by name. How, then, are we to characterize the relationship between these two *hay'a* works by Shīrāzī?

One obvious aspect of the relationship between these *hay'a* works has already been mentioned several times: the fact that these works were a record of Shīrāzī's changing views on the configuration of the heavens during the course of a short but particularly productive period. If nothing, else the fact that *Nihāya* and the *Ikhtīyārāt* were completed in such close succession, and that they differ in the details of how Shīrāzī presents his models, allows us to conclude that Shīrāzī's views on *hay'a* were subject to revision during the period in question (i.e., the period from 1281 C.E. to 1282 C.E., and indeed all the way to 1285 C.E. if we include the *Tuḥfa*).

Indeed, what we have seen of the *Nihāya* and the *Ikhtīyārāt* reflects an observation made by Ragep in the introduction to his critical edition of Ṭūsī's *Tadhkira*. Noting the abundance of commentaries that were written on Ṭūsī's *Tadhkira*, he

[61] Saliba, "Arabic Planetary Theories after the 11th Century AD," 99. Tellingly, in *Ikhtīyārāt-i Muẓaffarī*, Ayasofya MS 2575 Shīrāzī has written near the end of his geometric proof, "Yet, I harbor further considerations in this regard." This line is missing in *Ikhtīyārāt-i Muẓaffarī* Majlis MS 6398.

remarks on the high quality of these works and the fact that they included "new solutions to the *ishkālāt* (difficulties) of astronomy as well as very interesting passages concerning the status of astronomy, the relation of theory and observation, the role of physics in astronomy, and other theoretical concerns."[62] In his book *Islamic Science and the Making of the European Renaissance*, Saliba also highlights the commentary genre as having provided the opportunity for their authors "to produce their own alternative theories and to record their own scientific insight," cautioning as well against the failure to "appreciate the novel ideas that were contained therein."[63] Saliba further notes the importance of the "cumulative work of commentaries" as permitting "a tradition of dialogue with earlier astronomers," and highlights the importance of both the Marāgha observatory, and of Shīrāzī's *Nihāya* and *Tuḥfa*, in this regard.[64] Even though the passages that we have studied from the *Ikhtīyārāt* have been culled from one or two chapters at most (and thus may not be representative of the entire work) what we have seen is suggestive of one way to describe the *Ikhtīyārāt*. This work could be provisionally viewed as a commentary; a commentary in Persian, that is, on the author's own *Nihāya*.

In recognizing the importance of the commentary genre, Saliba considers the *Nihāya* as "one of the most elaborate Arabic *Hay'a* texts."[65] The *Ikhtīyārāt* allowed Shīrāzī to elaborate on several of the ideas that he merely hints at in the *Nihāya*. The most notable examples of this that we have encountered are Shīrāzī's discussion of the "four-fold hypothesis," which is merely mentioned in the *Nihāya*, as well as the discussion of the alignment point for which the reader of the *Nihāya* is expressly referred to the *Ikhtīyārāt*. The potential discovery of other elaborative sections in the *Ikhtīyārāt* will have to await the publication of the full texts of the *Nihāya* and the *Ikhtīyārāt*. Meantime descriptions of the *Ikhtīyārāt* (and by extension other Persian works on *hay'a*) as being secondary for the same qualities with which one could evaluate the *Nihāya* as "elaborate" is patently unfair (or, in the Arabic which Shīrāzī, Ṭūsī, and al-'Urḍī all used and loved, an instance of *al-kayl bi mikyālayn*).

We now look briefly at the other Persian texts mentioned in this chapter: Ṭūsī's *al-Risāla al-Mu'īnīya* (completed in 1235 C.E.), and *Ḥall-i Mushkilāt-i Mu'īnīya*. As was the case with the *Ikhtīyārāt* these are presented in "Persian Scientists in the Islamic World" to support Saliba's claim that the *hay'a* astronomers of the Islamic world chose Arabic exclusively as the vehicle for their important contributions to the field. Saliba considers the first of these works as elementary, when compared to Ṭūsī's later text in Arabic, the *Tadhkira*.[66] The key to Saliba's argument, however, is

[62]Ṭūsī, *Naṣīr al-Dīn al-Ṭūsī's Memoir*, 59.

[63]Saliba, *Islamic Science and the Making of the European Renaissance*, 240.

[64]Saliba, *Islamic Science and the Making of the European Renaissance*, 244.

[65]Saliba, "Persian scientists in the Islamic world," 141.

[66]Saliba, "Persian scientists in the Islamic world," 140.

the date for the second Persian text, the *Ḥall-i Mushkilāt-i Mu'īnīya*, in which Ṭūsī presents the planar version of his famed formulation, the Ṭūsī couple.[67] The dating for the *Ḥall-i Mushkilāt-i Mu'īnīya* is uncertain as none of the known surviving manuscripts list the date of completion for this work. According to Saliba this work was written "some time after 1247."[68] If this dating is accepted, then this would mean that Ṭūsī presented his couple in his redaction of the *Almagest*, the *Taḥrīr* al-Majistī (written in 1247) prior to including it in the *Ḥall-i Mushkilāt-i Mu'īnīya*. Writing on the controversy surrounding the dating for the *Ḥall-i Mushkilāt-i Mu'īnīya*, Ragep proposes a much earlier date, stating that it was probably written a few months after the *Risāla al-Mu'īnīya*.[69] Among Ragep's arguments for the early date of the *Ḥall-i Mushkilāt-i Mu'īnīya* are the fact that Ṭūsī does not mention the curvilinear version of the Ṭūsī couple in the *Ḥall-i Mushkilāt-i Mu'īnīya* (though he cryptically describes this in the *Taḥrīr* al-Majistī) and thus likely had not yet discovered it.[70] Though the matter of the dating has yet to be solved conclusively, it would be prudent – based on our experience with the *Ikhtīyārāt* – to not accept a late date for the *Ḥall-i Mushkilāt-i Mu'īnīya* based on an *a priori* assumption that Ṭūsī could not have consigned a description of his couple in Persian before doing so in Arabic. That approach would surely force a preconceived theory on the historical data, whereas historical research should ideally work in the reverse fashion.

Restricting ourselves to the findings in this chapter and the previous one, we can say with conviction that the *Nihāya* and the *Ikhtīyārāt* are a pair of closely related works composed during a period in which the author was rethinking and revising his models for the configurations of the universe. These books form a distinctive pair, as one is written in Arabic and the other, in close succession, in Persian, yet they share many of the same aims and the same scope, and they assume the same level of proficiency in their readers. The reputation of the *Ikhtīyārāt* as an abridgment or a translation of the *Nihāya* is undeserved and does not reflect that actual content of these two important *hay'a* works.

---

[67] Saliba, "Persian scientists in the Islamic world," 141.

[68] Saliba, "Persian scientists in the Islamic world," 141.

[69] Ragep, "The Persian Context of the Ṭūsī Context," 119. See also Ragep, "Ibn al-Haytham and Eudoxus: The Revival of Homocentric Modeling in Islam," 786–809; and Ragep and Hashemipour., "Juft-i Ṭūsī (the Ṭūsī Couple)," 472–475.

[70] Ragep, "The Persian Context of the Ṭūsī Couple," 119. Professor Ragep states "There is nothing in the *Ḥall* that was not promised to his patron Mu'īn al-Dīn in the *Mu'īnīyya*; in particular, he presents his solution for the Moon and planets using the rectilinear version of his couple . . . , and, most significantly, he does not offer any solution of his own for the Moon's prosneusis problem nor for the planetary latitude problem, which he much later solved with his curvilinear version of the couple...."

5.4 Conclusion

Among other things, this chapter has been concerned with the meaning of the word Persian. Recalling that the word Persian can be used in several guises, we should ask: On what sense of the word Persian should we rely when conducting historical research on the medieval scientists of the Islamic world? Do we mean to use this word as a reference to the racial background of the scientists (which as Saliba suggests is generally not subject to verification and therefore untenable)? Do we rather take this word to refer to the geographical location in which the author wrote his work?[71] Or do we perhaps refer to the shared cultural histories of the Persianate lands in western Asia? As a historian of the Middle East, I confess my preference for the latter, with the conviction that there is much to be lost – and not much gained – in treating the various constituent cultures of the vast lands of Islam (and their corpora of texts) as a monolithic entity. To be sure, when the scientists of medieval Islam themselves referred to the scholars of the Persian-speaking world (regardless of whether or not they identified themselves as belonging to this group) they were often keen to emphasize the fact that these scientists were Muslims or "Arab in their religion."[72] But the same authors often make an effort in describing the Persian-speaking scholars of Islam through the lens of an Arab vs. *'ajam* dichotomy that is rather persistent, and that should be accounted for when carrying out research on the rich textual traditions of the Islamic world.[73]

Given the ubiquity of Arabic as a language of learned discourse in the Islamic world, one has to ask furthermore, how sensible it is to try and study the cultural influence of Persian speaking scientists (in *'Irāq-i 'ajam*, Khurāsān, Transoxiana, and elsewhere) through those of their works that they wrote in Persian? Or, to view the question slightly differently, would it be as tenable for an inquiry concerning the life and the scholarly output of Copernicus or Newton (or, even, Buridan) if one were to categorize these authors as "Latin," by virtue of the language in which they chose to communicate their scientific works? In his book *Cotton, Climate and Camels*, Bulliet describes a reductive historiographical trend by which the "historical moment" of the Persianate world is "elided with that of the Arab Muslims whose extraordinary conquests had brought Iran into the caliphal empire."[74] He adds "specialist on matters Arabian frequently forget to mention how many of the most prominent authors of medieval works in Arabic grew up in Persian-speaking homes."[75] Works such as the *Ikhtīyārāt* are important because, they offer a direct

---

[71] As did Kennedy in his survey of the *zīj* literature. See note 4.

[72] See Bīrūnī, *Āl-Biruniś Book on Pharmacy and Materia Medica*, 13, and Ibn Khaldūn, *The Muqaddima*, 543.

[73] See Bīrūnī, *Chronology of ancient nations; an English version of the Arabic text of the Athâr-ul-Bâkiya of Albîrûnî*, 226, and Ibn Khaldūn, *The Muqaddima*, 543.

[74] Bulliet, *Cotton, Climate, and Camels in Early Islamic Iran* (New York: Columbia University Press, 2009), 128.

[75] Bulliet, *Cotton, Climate, and Camels in Early Islamic Iran*, 128.

window into the realm of this Persianate culture within Islam. Indeed, our foregoing discussions regarding the nature of the *Ikhtīyārāt* and how this astronomical work has been misrepresented in the literature indicates that by ignoring the Persianate identity of the scientists of Persia, or by "eliding" the Persianate presence within Islam with the Arabic one, we risk a skewing of the framework within which carry out our inquiry, thus misunderstanding the nature of the scientific works produces in the Perso-Islamic world, during the medieval period.

Chapter 6
Concluding Remarks

6.1 The Interrelation of Shīrāzī's Works on *Hay'a*

The similarities between the outlines of the *Nihāya* and the *Tuḥfa* (and the fact that these are traceable to Ṭūsī's *Tadhkira*) have been known to historians for many decades. This study indicates that the *Ikhtīyārāt* shares this feature with its two companions. The fact that these three works follow the same essential format but contain formulations that differ in their detail, suggests that they are, among other things, a record of Shīrāzī's changing views on *hay'a*. This can be seen in the variations in planetary configurations as they appear in each of the works. Taken as a whole these texts seem each to have afforded Shīrāzī the opportunity to revise and rework his astronomical theories. As such these works provide a rather rare opportunity to follow the creative process of an important *hay'a* author of the late thirteenth century.

Though limited to a small section of the aforementioned astronomical works by Shīrāzī, this study has further uncovered three unexpected features in these texts. The first has to do with the *Ikhtīyārāt*. It is clear that the date for the *Ikhtīyārāt* is sandwiched between the date for the other two works studied here (indeed, as we saw, the *Ikhtīyārāt* selections was published a mere 4 months after the *Nihāya*). Thus, Shīrāzī completed all three of his major works on astronomy in a little under 4 years, i.e., from 1281 to 1285 C. E. This was during a fecund if at times hectic period in Shīrāzī's career, prior to what he dubs a period of cataclysms. As we saw in Chap. 3, this period of intense productivity (which involved the completion of a number of works on topics other than astronomy) appears to have been followed by a fallow period that lasted for more than a decade.

The second remarkable fact in regard to Shīrāzī's works and their dates of publication is the manner in which Shīrāzī's revisions and reworking were captured in his books. We have seen that the astronomical models for the superior planets that appear in the *Ikhtīyārāt* correspond to the earliest versions of these as they appear in

K. Niazi, *Quṭb al-Dīn Shīrāzī and the Configuration of the Heavens: A Comparison of Texts and Models*, Archimedes 35, DOI 10.1007/978-94-007-6999-1_6,

the *Nihāya*. It appears that while the text of the *Nihāya* was subsequently amended these changes did not end up in the *Ikhtīyārāt*.[1]

The extensive revisions made by the author in the *Nihāya* are themselves, of course, the third remarkable fact with respect to the chronology of these works and their relationship with each other. These appear primarily in the earliest known manuscript of this work, MS Köprülü 957. Many of these revisions appear in Shīrāzī's discussion regarding the orientation (or the tilt relative to the plane of the ecliptic) of the orbs for the superior planets, but are not limited to this section. The corrections make it rather clear that, contrary to Shīrāzī's claim as to the status of the *Nihāya* as a mature work, he subjected the astronomical theory that was included in the *Nihāya* to heavy revisions. As we have seen, some of these revisions were due to the fact that Shīrāzī did not view all of his proposed models in the *Nihāya* and the *Ikhtīyārāt* as satisfactory. He was therefore forced to abandon some of the models which he had proposed in these works, when authoring his third book, the *Tuḥfa*. As a result one of the only references in the *Tuḥfa* to Shīrāzī's earlier models of the superior planets occurs in the chapter on the hypotheses,[2] where the author obliquely refers to some of his earlier models as untenable on observational grounds, while claiming that these were included in the earlier works as a test of the intelligence of the reader.

6.2 Physical and Mathematical Principles in Shīrāzī's *hay'a*

As far as Shīrāzī's theoretical approach, our work in Chaps. 4 and 5 has highlighted two interrelated and ever-present themes. The first has to do with the central important of the "hypotheses." To Shīrāzī these hypotheses were mathematical formulations representing what was in effect a sum of vectors moving with uniform angular motions (i.e., the system referred to as one of "wheels upon wheels"), and which – with the judicious choice of vector and of angular motion – could be configured to match the observed behavior of the heavenly bodies. We have noted Shīrāzī's debt to Ṭūsī for his chapter on the hypotheses,[3] while noting as well that Shīrāzī's chapters on the hypothesis in the *Nihāya* and the *Tuḥfa* are expanded relative to the material in Ṭūsī's chapter on the hypotheses in the *Tadhkira*; for they include hypotheses that were either not included in the *Tadhkira* (such as 'Urḍī's Lemma), or those that Ṭūsī included in other parts of his book (such as the Ṭūsī couple in its planar and spherical variations, see Appendix D). Shīrāzī's desire to

[1]The reasons for why the heavy emendations did not find their way into the *Ikhtīyārāt* are not clear, but are perhaps related Shīrāzī's claims as to the status of the *Nihāya* as his seminal work (i.e., one in which he was willing to lavish time and effort upon subsequent to its publication). The fact that these emendations do not appear in the *Ikhtīyārāt* may also have been because this work existed in some form prior to the *Nihāya*, as will be discussed in the next section.

[2]For its full confirmation this conclusion will have to wait for a complete edition of the *Tuḥfa*.

[3]Ṭūsī, *Naṣīr al-Dīn al-Ṭūsī's Memoir*, x.

provide in this chapter as comprehensive a set of hypotheses as he was capable, underscores his approach to his astronomical research, which appears to have been aimed at ridding his planetary models of their perceived faults with the aid of a full battery of hypotheses.

In Chap. 4 we also noted how the *Ikhtīyārāt* chapter on the hypotheses is unusual in that in this work, rather than provide a full list of the hypotheses as these appear in the *Nihāya*, Shīrāzī only includes four while presenting the remaining hypotheses in other parts of his book (see Appendix D). The four hypotheses that Shīrāzī does include are precisely the ones included by Ṭūsī in his chapter on the hypotheses in the *Tadhkira*.[4] The reason for this is not clear, though this arrangement suggests that what appears in the *Ikhtīyārāt,* may be the reflection of an earlier conception for the presentation of the hypotheses, i.e., one predating the *Nihāya*. If future research were to confirm this theory in regard to Shīrāzī's manner of presentation of the hypotheses in the *Ikhtīyārāt*, then this would indicate that Shīrāzī's earliest thinking with respect to the hypotheses was preserved by him in some form in his native Persian. It would then make sense that for this chapter of the *Ikhtīyārāt* Shīrāzī decided to use what was at hand rather than carrying out the additional work of translating or commenting on *Nihāya* material into Persian. Though the presence of this Persian urtext is conjectural, one can imagine how the arrival of Muẓaffar al-Dīn at the Mongol court could have provided Shīrāzī with the opportunity to dust off (and perhaps to revise) work that he had completed earlier, and to use it for cultivating a new patron-client relationship.

As a final note on Shīrāzī's conceptualization of the hypotheses we emphasize again his awareness of the importance of the works of his predecessors, especially in regard to several non-Ptolemaic hypotheses that were able to effectively challenge the models of Ptolemy, and his urge to compile these hypotheses. It is important, for instance, to note that 'Urḍī's Lemma did not appear in the works of Ṭūsī, nor did the Ṭūsī couple find its way in 'Urḍī's *Kitāb al-hay'a.* Shīrāzī's ultimate goal was to use his comprehensive mathematical (and conceptual) toolkit for the resolution of the *ishkālāt* in as methodical a manner as possible. To be sure, in following this approach Shīrāzī was operating within a longstanding tradition of *hay'a*, but in his focus on the compilation and the tabulation of these hypotheses (with each hypothesis earmarked for specific planetary anomalies), he seems to have privileged these formulations even more than they had been previously.

The second underlying theme of Shīrāzī's astronomical theory is one that it also shares with other works of the *hay'a* genre, and this is the importance of the laws of physics, and the importance of consistency between the hypotheses and these laws.[5]

[4]Ṭūsī, *Naṣīr al-Dīn al-Ṭūsī's Memoir*, 130–143; Shīrāzī, *Ikhtīyārāt-i Muẓaffarī*, Ayasofya MS 2575, 57r – 65v.

[5]Saliba, "Aristotelian Cosmology and Arabic Astronomy," 253–257. Shīrāzī's novel "physical law" in regard to the *prosneusis* point, sketched out in Chap. 5, is particularly relevant here, however. Once studied fully it will likely result in a re-evaluation of what Shīrāzī (and those working in the same *hay'a* tradition) considered to be a law of natural philosophy. Certainly, in posing it Shīrāzī appears to have abandoned the need to remain consistent with solid geometry.

At the beginning of his chapter on the hypotheses Shīrāzī states: "So we say that motions that are non-uniform, as apparent from observation, and which may not issue from the celestial orbs except due to a displacement [i.e., of the observer from the center as in an eccentric orb] or a combination of uniform motions [that in turn necessitate non-uniformity with respect to us, i.e., the observer] – occur in several varieties."[6] This stipulation of uniform circular motion, which was effectively the cornerstone of the *hay'a* genre as a whole, echoes a statement by Ṭūsī in his chapter on the *uṣūl* in the *Tadhkira*,[7] and is ultimately due to Ptolemy.[8] It is therefore important to note, as well – as an indication of the self-conception of scientists of the *hay'a* tradition – the manner in which Shīrāzī emphasizes Ptolemy's lack of conformity to the laws governing the motion of solid spheres.[9] In the *Ikhtīyārāt* he writes: "And as for Ptolemy, who is the founder of the principles (*qawā'id*) and the master of observation, since he doesn't account for solid bodies, and is content rather with presenting lines and circles according to his own goals, he is exempt from this task [i.e., to provide a coherent explanation of the configuration of the orbs while remaining in agreement with observation and the principles of astronomy]."[10] We see here the driving force of Shīrāzī's *hay'a* research, and of *hay'a* research in general; to take the cosmological system inherited from Ptolemy and to re-work it in a framework of nested spheres that would be true to a set of assumptions, which were consistent (generally speaking) with the physics of uniformly rotating spheres.

[6]Shīrāzī, *Nihāyat al-idrāk fī dirāyat al-aflāk*, Köprülü MS 956, 33v.

[7]"If a celestial motion is irregular from our perspective, we must require that it have a [hypothesis] according to which that motion is uniform; this [hypothesis] should also bring about its irregularity with respect to us. For irregular [motion] does not arise from the celestial bodies." Ṭūsī, *Naṣīr al-Dīn al-Ṭūsī's Memoir*, 130.

[8]Ptolemy, *The Almagest*, 141.

[9]In the chapter on the hypotheses in the *Ikhtīyārāt* Shīrāzī states:

> "So it is incumbent upon the group of moderns, who talk about [the corporeality] of the orbs and the descriptions of the principles of the motions that they have obtained through observation [while?] they establish an orb that acts as mover for each motion – [and this group of moderns includes] Abū 'Alī ibn al-Haytham who was a prominent mathematician, whose words and words of others like him have greatly [influenced] the configuration of the orbs as three-dimensional bodies – to describe the [configuration of the] orbs in a manner such that that which is desired is obtained from it, while at the same time it is consistent with the principles [of hay'a]. And should [the account] add or subtract from the number of orbs it will not be [an issue] but if it is inconsistent with what is found through observation or if is not [consistent] with some of the rules and [principles] then it will have [missed its mark]." Shīrāzī, *Ikhtīyārāt-i Muẓaffarī*, Ayasofya MS 2575, 67v. The description of Ibn al-Haytham is quoted directly from *Ḥall-i mushkilāt-i mu'īnīyah*. See Ṭūsī, *Ḥall-i mushkilāt-i mu'īnīyah*, 14.

[10]Shīrāzī, *Ikhtīyārāt-i Muẓaffarī*, Ayasofya MS 2575, 68r. See note 45, Chap. 5.

As we have seen, the fact that Shīrāzī opted not to use 'Urḍī's Lemma in the model for the superior planets for the *Ikhtīyārāt* has only been brought to light in the last several years.[11] One of the findings of the present study is that Shīrāzī decided against the use of 'Urḍī's Lemma in the *Nihāya*, as well; favoring instead – as he did in the *Ikhtīyārāt* – his own rendition of Apollonius's theorem, the Conjectural Hypothesis. As we saw in Chap. 5 the reason for this may have been related to Shīrāzī's desire to maintain a perfectly circular path for the center of the epicycle.[12] It is clear, at any rate, that Shīrāzī's was fully aware of 'Urḍī's work while writing the *Nihāya,*[13] and that his decision to propose an original model for the superior planets was not due to his ignorance of 'Urḍī's model for the superior planets. Indeed, since this choice was clearly due to Shīrāzī's confidence in his own ability to do better, the *Tuḥfa*, would have provided – among other things an opportunity – for Shīrāzī to concede (if obliquely) the untenability of his own model for the superior planets.

Looking beyond Shīrāzī's model for the superior planets as these appears in the *Nihāya* (and the variants in the *Ikhtīyārāt*), we note here Shīrāzī's contribution in applying 'Urḍī's formulation to what was an original configuration for the Moon. This model successfully addressed one of the issues with Ptolemy's proposed configuration in that it allowed the motion of the center of the Moon's epicycle to be described as a combination of circular motions about the center of the universe.[14] It is also worth repeating here that the present study has demonstrated that the use of 'Urḍī's Lemma for the configuration of the Moon should not be considered an innovation that appears in the *Tuḥfa* only, but that it was already included by Shīrāzī in his earlier work, the *Nihāya*.[15] In the *Nihāya*, Shīrāzī states in regard to al-'Urḍī's failure to recognize the importance of his own formation for the configuration of the Moon: "And the master of this principle did not [recognize its (i.e., the principle's) application] in proving the uniformity of the motion of the center of the Moon's epicycle about the center of the universe as [we have recognized] and for this reason he [took refuge] in proving this via [reversing] the directions of motion [i.e., of the deferent and the encompasser spheres]."[16] Given the ringing self-endorsement in regard to the configuration of the Moon, Shīrāzī's praise for al-'Urḍī by name

---

[11]Gamini, "The Planetary Models of Quṭb al-Dīn Shīrāzī in the *Ikhtīyārāt-i Muẓaffarī*."

[12]See note 39, Chap. 5. The circularity of the orbit appears to have been important to Shīrāzī because of his ideas in regard to the alignment point, as we saw in Chap. 5.

[13]He quotes *Kitāb al-hay'a* both in the chapter on the Moon, and in his discussion of 'Urḍī's Lemma.

[14]Saliba, "Arabic Planetary Theories after the 11th Century AD," 99.

[15]Shīrāzī, *Nihāyat al-idrāk fī dirāyat al-aflāk*, Köprülü MS 957, 95r.; See also Saliba, "Arabic Planetary Theories after the 11th Century AD," 97–98.

[16]Shīrāzī, *Nihāyat al-idrāk fī dirāyat al-aflāk*, Köprülü MS 957, 95r.; For a discussion of Shīrāzī's claim in the Tuḥfa to having solved the alignment issue see Saliba, "Arabic Planetary Theories after the 11th Century AD," 99.

in the late work the *Durra*, therefore may be viewed, again, as a sign of respect and admiration for his predecessor, as well as an implicit concession to the relative excellence of ʻUrḍī's model for the superior planets.[17]

6.3 Alchemy at Marāgha

We have noted earlier the intriguing association of the Marāgha observatory with alchemy. As was pointed out this association was already noted by Sayılı in his book *The Islamic Observatory*. In discussing the same passages from Rashīd al-Dīn's narrative of Hülegü that we reviewed in Chap. 2, Sayılı states:

> The same author [i.e., Rashīd al-Dīn] tells us that Hulāgū allotted salaries and pensions to the scientists and philosophers and had his royal residence embellished with their presence. The emphasis here seems to be on pseudo-sciences such as astrology and alchemy. Indeed, there is ample evidence concerning the astrological side of that interest, and Rashīd al-Dīn informs us that Hulāgū had a special inclination toward alchemy and dwells at some length on his wasted confidence on the alchemists. He says that they kindled much fire, constructed many a vessel, employed bellows of various sizes and consumed immeasurable amounts of materials but that although they caused the expenditure of immense sums of money they did not produce a particle of silver or gold and it all came to naught and resulted in no benefit to anyone except that these impostors thereby secured a livelihood for themselves. It seems probable therefore that Marāgha was also the scene of alchemical activities of considerable extent.[18]

Sayılı's main concern in this passage from which the fragment is excerpted is a discussion of Marāgha as a locus of contact between the astronomical traditions of the Islamic world and the Far East, and so he does not stress the significance of Rashīd al-Dīn's comments on Marāgha as a site of alchemy.[19] What Sayılı appears to overlook, and which is clarified only upon a comparison of the narrative of Hülegü's death with that of his grandson Arghūn, is that one of the main reasons for alchemical research at Marāgha could very well have been to grant longevity or immortality to the Ilkhanid ruler. This is suggested by the fragment of Rashīd al-Dīn describing Arghūn's death (see Chap. 2, Sect. 2.3.4). If we accept this, it then follows that – rather than focusing on the transformation of metals – the alchemy practiced at Marāgha may have had a close affinity Taoist or other "eastern" traditions with a strong interest in the elixir of immortality, and its use for granting immortality to the ruler.[20]

---

[17]Majlis MS *Durrat al-Taj* 4729, 121r. See note 47, Chap. 1.

[18]Sayılı, *The Observatory in Islam*, 193.

[19]Sayılı, *The Observatory in Islam*, 192. At any rate, Sayılı's reference to astrology and alchemy as "pseudo-sciences" is anachronistic. Certainly to Ilkhanid ruler and subject alike, these activities were scientifically sound (at least as far as their epistemological validity).

[20]J. C. (Jean C.) Cooper, *Chinese Alchemy: The Taoist Quest for Immortality* (Wellingborough, Northamptonshire: Aquarian Press, 1984), 19; See also Zhichang Li, *The Travels of an Alchemist;*

According to Rashīd al-Dīn, the adept who prepared and administered this elixir to Arghūn "came from India."[21] According to a chapter in Bīrūnī's *India* entitled "في ذكر علوم لهم كاسرة الأجنحة على أفق الجهل" – translated as "Of Hindu Sciences Which Prey on the Ignorance of People" by Sachau[22] – Indian alchemy appears to have included a tradition called the *Rasayana*, which, like alchemy in the Taoist tradition, was principally concerned with the rejuvenation of the vital spirit.[23] This tradition was primarily based on herbal preparations rather than metallic ones, however, at least according to Bīrūnī,[24] so it is likely unrelated to alchemy as practiced at Marāgha. It is certainly possible that the adept who attended Arghūn in Rashīd al-Dīn's account (which was written several decades after the events they describe) subscribed to a tradition different from *Rasayana.* It is also possible, however, that Rashīd al-Dīn's adept was from China or some other place in the Far East, and that the account of his origin were distorted. That Arghūn's adept followed Taoist beliefs or a related system is suggested, however, by the fact the text describing Arghūn's seclusion, and his partaking of the alchemical draught bears a striking similarity to a description of the Taoist tradition of "Potable Gold elixir" meant to have life-prolonging qualities. This elixir was apparently based on minerals and metals, rather than herbal potions. Commenting on the difficulties in obtaining it, the Taoist author Ko Hung (or Gě Hóng, 283–343 C.E.) describes the following regimen as its prerequisites: "money, seclusion in some famous mountain-range, isolation from profane unbelievers and critics, religious ceremonies, purificatory rites; abstention from pungent flavours and fish, to say nothing of the fasting; long heating under exact condition of temperature, needing taxing watch; and finally the indispensability of oral instruction from a genuine adept, as teacher."[25] As can be seen several of these elements occur in Rashīd al-Dīn's account of Arghūn's death; most notable among them the need for seclusion, the importance of purificatory rites, and of the constant accompaniment of the adept/guide. The exact tradition upon which Arghūn relied in his quest for immortality is perhaps less important, however, than the implication that at least for part of its existence Marāgha was

the Journey of the Taoist, Ch'ang-Ch'un, from China to the Hindukush at the Summons of Chingiz Khan, Recorded by His Disciple, Li Chih-Ch'ang (London: G. Routledge & sons, Ltd., 1931).

[21]Rashīd al-Dīn Ṭabīb, *Jāmi' al-tawārīkh*, 824. See Chap. 2, Sect. 2.3.4 for a translation of this passage. See also note 131 in the same section.

[22]A rough translation of this expression could be rendered as follows: "On those sciences that entail a folding of the wings (i.e., an alighting) upon the horizon of ignorance."

[23]Muḥammad ibn Aḥmad Bīrūnī, *Kitāb fī taḥqīq mā lil-Hind min maqbūlah fī al-'aql aw mardhūlah (Hyderabad: Osmania Oriental Publications Bureau, 1958), 150; Muḥammad ibn Aḥmad Bīrūnī, Alberuni's India. An Account of the Religion, Philosophy, Literature, Geography, Chronology, Astronomy, Customs, Laws and Astrology of India About A.D. 1030*, (London: K. Paul, Trench, Trübner & Co., Ltd., 1910), vols. 1, 188.

[24]Bīrūnī, *Kitāb fī taḥqīq mā lil-Hind min maqbūlah fī al-'aql aw mardhūlah*, 150.

[25]Joseph Needham, *Science and Civilisation in China* (Cambridge [Eng.]: University Press, 1954), vols. 5, part 2, 68.

involved with alchemical technology meant to prolong the Ilkhanid ruler's life.[26] Certainly the very possibility should allow us to view the strategic importance of Marāgha to the well-being of the Ilkhanid polity in a different light.

6.4 Persian vs. Arabic in the Scientific Works of the Persianate World: Future Work

What we have seen previously in regard to the *Ikhtīyārāt* indicates that this book does not fit the description of a derivative or simplified *hay'a* work written in Persian. Rather, the technical sophistication of the *Ikhtīyārāt* matches that of Shīrāzī's other *hay'a* works, i.e., the *Nihāya* and the *Tuḥfa*.[27] This study underscores the fact, therefore, that the prevalence of Arabic works in *hay'a* over those written in Persian, should not be mistaken as an *a priori* indication of mediocrity in the latter. We have touched earlier on the reasons for the prestige and dominance of Arabic as a language of scholarship. Here we revisit a telling bit of evidence from Shīrāzī's works examined above. The fact that in the *Nihāya* (i.e., in what was nominally his seminal work) Shīrāzī refers his readers to the Persian text of the *Ikhtīyārāt* for an exposition of a technical point, can only mean that Shīrāzī expected the readers of his *Nihāya* to have the ability to a read technical *hay'a* texts in Persian, as well. This fact raises a number of questions in regard to the *hay'a* texts of Shīrāzī, Ṭūsī, and their colleagues, and their intended audiences. Were these works, for example, meant to be read by readers across the Islamic world or were they written with a more limited readership in mind? There is no way to answer this conclusively without actually studying a representative sample of the Persian *hay'a* texts, as well as a representative sample of Arabic *hay'a* texts produced in the Persianate world.

As we saw, the great fourteenth century historian Ibn Khaldūn viewed the Mongol invasions as having virtually snuffed out the cultural productivity of Persian-speaking lands. Yet, this evaluation stands in contrast to the work of Shīrāzī and his predecessors in the Persianate world. In his essay entitled "The Exact Sciences," (in the fifth volume of *The Cambridge History of Iran*), Kennedy dubs the scientists of the Saljuq and Mongol Iran as "the best of their age,"[28] and Saliba identifies the thirteenth century as one in which scientific production of the Islamic world continued to flourish.[29] Certainly, what bears directly on this paradox is the vast number of existing manuscripts, in Arabic certainly, but also in Persian, that

---

[26]Li, *The Travels of an Alchemist*, 113.

[27]This is contrary to Professor Saliba's evaluation of this work. See Saliba, "Persian scientists in the Islamic world," 141.

[28]Kennedy, "The Exact Sciences in Iran Under the Saljuqs and Mongols," 679.

[29]Saliba, *Islamic Science and the Making of the European Renaissance*, 236.

remain to be studied.[30] Ultimately, a considerable amount of work remains to be done in order to establish the manner both in which the Mongols affected the cultural and scientific lives of their subjects, as well as the way in which Shīrāzī and his fellows scientists negotiated their cultural backgrounds as Persians, both with respect to the Mongol ruling class, as well as relative to the tradition of Islamic scholarship as a whole. In rejecting the standard narrative of a post-Ghazālī intellectual "decline" in the Islamic world, Saliba traces a flourishing tradition of astronomical research that continues through the era of Shīrāzī, and from there to the fifteenth century and beyond.[31] The continuity and perseverance of this scientific tradition and its relevance to the Early Modern Era only serve to underscore the importance of studying the scientific works of Shīrāzī – and that of his colleagues in Persia and elsewhere in the Islamic world who were the best of their age.

[30]Given our present state of knowledge it is inevitable that some our conclusions in regard to the role of scientific works written in Persian will be in need of revision in the future. A look at Storey's bibliographic survey of Persian literature demonstrates that the Persian manuscripts represent on their own, a considerable body of material that has been languishing for lack of attention. Storey, *Persian Literature*, vol. 2, 1:35–117.

[31]Saliba, *Islamic Science and the Making of the European Renaissance*, 240.

Appendices

Appendix A

Table of Contents, The *Nihāya*

K. Niazi, *Quṭb al-Dīn Shīrāzī and the Configuration of the Heavens: A Comparison of Texts and Models*, Archimedes 35, DOI 10.1007/978-94-007-6999-1,

Chapter 12. On the variation in the moon's illumination and on lunar and solar eclipses, and the time period between two lunar or solar eclipses.
Chapter 13. On sectors and the situation of visibility and invisibility and conjunctions.

Book III: On the configuration of the earth and its populated and desolated sectors and the [consequences] accruing to it due to the changing positions of the celestial bodies.

Chapter 1. A general summary of the configuration and circumstances of the Earth.
Chapter 2. On the characteristics of the equator.
Chapter 3. On the characteristics of locations having latitude which are called the oblique horizons, and on the extent of east and west and the equation of daylight.
Chapter 4. On the characteristics of locations whose latitude does not exceed the complement of the obliquity.
Chapter 5. On the characteristics of locations whose latitude exceeds the complement of the obliquity but does not reach one-quarter revolution.
Chapter 6. On the characteristics of locations whose latitude is exactly one-quarter revolution.
Chapter 7. On the co-ascensions of the ecliptic.
Chapter 8. On the lengths of the nychthemerons.
Chapter 9. On dawn and dusk.
Chapter 10. On understanding the units of the day, namely hours.
Chapter 11. On the degrees of transit of the stars on the meridian and on their [degrees] of rising and setting.
Chapter 12. On shadows.
Chapter 13. On the meridian line and the direction of the *qibla*.

Book IV: On finding the measurements of the distances and the bodies.

Chapter 1. Introduction
Chapter 2. On the measure of the Earth, and on the knowledge of the height of the sphere of air.
Chapter 3. On finding the distances of the Moon from the center of the world.
Chapter 4. On the sizes of the diameters of the Moon.
Chapter 5. On the measure of the diameter of the Sun and the ratio of the sizes of the two luminaries and the Earth.
Chapter 6. On the other dimensions of the Sun and the dimensions of the two lower bodies [and their size].
Chapter 7. On the measure of the superior planets.
Chapter 8. On the distance of the fixed stars and their bodies:
Chapter 9. On the erroneous distances and sizes due to early and late astronomers without exception.
Chapter 10. On the correct method for extracting distances and sizes.

Table of Contents, The *Tuḥfa*

Book I: On introductory remarks that need to be made prior to commencing on our desired discussions and that is in three chapters.

Book II: On the configuration of the celestial bodies and related topics, of the relationships between some of the bodies in thirteen subsections.

Book III. On the configurations of the Earth generally [lit. whether filled or empty] and all that is properly related to it, in view of the differences with those of the superior planets. This is in thirteen chapters.

Chapter 1. On the configuration of the Earth.
Chapter 2. On the properties of the equator.
Chapter 3. On the properties of locations with finite latitude and these are called oblique horizons, and on the extent of east and the west and the equation of daylight.
Chapter 4. On the properties of locations where the latitude does not cross total obliquity.
Chapter 5. On the properties of locations where the latitude crosses total obliquity but does not reach a quarter of revolution.
Chapter 6. On the properties of locations where the latitude is a quarter of revolution.
Chapter 7. On the zodiacal co-ascensions.
Chapter 8. On the angles of the planetary transits and the angles of their rising and setting.
Chapter 9. On the length of day and night and the day and night equations.
Chapter 10. On morning and dawn.
Chapter 11. On the divisions of the day, i.e., the hours, and on what are composed of days such as months, years and related topics such as leap years and dating.
Chapter 12. On shadows.
Chapter 13. On the meridian line and on the direction of the *qibla*.

Book IV. On the distance and size of the planets in three chapters.

Chapter 1. On distances and sizes as they are commonly understood and this is in three articles and two principles, the first article on introductory remarks that are needed prior to commencing on our desired discussion, the second article on the area of the earth and what is properly related to it, the third article on the determination of the unknown sides/ angles in a triangle? from the known.
Chapter 2. On the demonstration of the error of the ancients and the moderns in the determination of sizes and distances.
Chapter 3. On the proper way to determine distances and sizes.

Table of Contents, *Ikhtīyārāt*

Book I: On introductory remarks that need to be made prior to commencing on our desired discussions and that is in three chapters.

Chapter 1. Definition of the science of planetary configurations, its subject matter, its foundations and issues in summary fashion.

Chapter 2. On introductory statements that belong to geometry, and consists of two articles; the first one devoted to definitions and the second one containing geometric theorems that are needed.
Chapter 3. On introductory remarks pertaining to the natural sciences in two chapters. The first on the classes of solid bodies and their motion in summary fashion and the second chapter on matters pertaining to the natural sciences.

Book II: On the configuration of the celestial bodies and related topics, of the relationships between some of the bodies in thirteen subsections.

Chapter 1. On the sphericity of the visible surface of the earth and the sphericity of the heavens as perceived by the senses and how the earth relative to the heavens is as the center of a sphere to its surface and how the earth is stationary at the center and this is in four sections. The first is the sphericity of the visible portion of the earth and water, the second on the sphericity of the heavens as perceived by the senses, the third on how the earth is unto the sky as is the center of the sphere to its surface and fourth on how the earth is stationary at the center.
Chapter 2. On the description of the simple bodies.
Chapter 3. On the well-known circles great and small.
Chapter 4. On the causes of the primary and secondary motions and the fixed stars.
Chapter 5. On accounting for *(isnād)* the motions that appear forbidden by the motion of the spheres such as fastest speed and slowest speed and retrograde motion and station based on hypotheses *(usūl)* that would permit their occurrence and on the configuration of the planetary spheres in summary fashion and a brief mention of the existing difficulties and it consists of four sections; first on a description of the cause of fastest and slowest speeds, second on a description of the cause of retrograde motion and station and direct motion, third on the ways in which the solid spheres can be envisioned and its mapping onto a planar surface and the realization of the flat figure, on the configuration of the spheres in the well-known manner and a brief mention of the difficulties that lie therein.
Chapter 6. On the spheres and the motions of the Sun.
Chapter 7. On the spheres of the Moon and its motion in longitude and latitude.
Chapter 8. On the spheres of the superior planets.
Chapter 9. On the spheres of Venus and Mercury and their motions in longitude.
Chapter 10. On the latitudes of the five planets who are called the wandering planets and this includes the purpose an introduction and a conclusion. As for the introduction it is an explanation of the situation of the apogees and nodes of these planets and the conclusion is an exposition of the spheres for the seven planets and in our reckoning these come out to forty-five, etc.
Chapter 11. On parallax.

Chapter 12. On the variation in the light from the Moon and solar and lunar eclipses and the period between two subsequent solar eclipses or lunar eclipses; and this includes an introduction and four articles and a conclusion. The introduction on conjunctions expresses how the position of the two luminaries is the same point on the ecliptic and its ascendant corresponds to the ascendant of the conjunction. The first article is on the variation of the luminosity of the moon; the second on the lunar eclipse; the third on the solar eclipse; the fourth on the period between successive lunar eclipses and solar eclipses, the conclusion on the planetary sectors, conjunctions, *tashrīq, taghrīb,* and visibility and invisibility *(khafā')*.

Book III. On the configurations of the Earth generally and all that is properly related to it, in view of the differences with those of the superior planets. This too is in thirteen chapters.

Chapter 1. On the configuration of the Earth, and a brief bit on its condition or state.
Chapter 2. On the properties of the equator.
Chapter 3. On the properties of locations with finite latitude and these are called oblique horizons, and on the extent of east and the west and the equation of daylight.
Chapter 4. On the properties of locations where the latitude does not cross total obliquity.
Chapter 5. On the properties of locations where the latitude crosses total obliquity but does not reach a quarter of revolution.
Chapter 6. On the properties of locations where the latitude is a quarter of revolution.
Chapter 7. On the zodiacal co-ascensions.
Chapter 8. On the angles of the planetary transits and the angles of their rising and setting.
Chapter 9. On the length of day and night and the day and night equations.
Chapter 10. On morning and dawn.
Chapter 11. On the divisions of the day, i.e., the hours, and on what are composed of days such as months, years and related topics such as leap years and dating.
Chapter 12. On shadows.
Chapter 13. On the meridian line which is also called the vanishing line and on the azimuth of cities.

Book IV. On the distance and size of the planets in three chapters.

Chapter 1. On distances and sizes as they are commonly understood and this is in three articles and two principles, the first article on parallel lines and introductory remarks that are needed prior to commencing on our desired discussion, the second article on the area of the earth and what is properly related to it, the third article on the determination of the unknown sides/angles

in a triangle from the known, the first principle on the determination of a more optimum method for the determination of sizes and distances and includes six rules, the second principle on the better-known method for the determination of sizes and distances which includes an introduction and five rules.
Chapter 2. On the demonstration of the error of the ancients and the moderns in the determination of sizes and distances.
Chapter 3. On the proper way to determine distances and sizes.

Appendix B: Chapter Outlines for the Chapter on the Superior Planets

نهاية الادراك في دراية الافلاك	اختيارات مظفرى	التحفة الشاهية
The Nihāya	*The Ikhtīyārāt*	*The Tuḥfa*
1. The observational basis for the configuration of the orbs.	**1. The observational basis for the configuration of the orbs.**	**1. The observational basis for the configuration of the orbs.**
2. The Orbs (Note: the model presented in this part of the Nihāya is a replica of what appears in the chapter on the Moon in Ṭūsī's *Tadhkira.*For a discussion of Shīrāzī's original model in the *Nihāya*see Chapter Four.)	**2. The Orbs**	**2. The Orbs**
2.a. The parecliptic	**2.a. The parecliptic**	**2.a. The parecliptic**
2.b. The Eccentric Deferent	**2.b. The Eccentric Deferent**	**2.b. The Eccentric Deferent**
	2.c. The Encompasser	**2.c. The Encompasser**
	2.d. The Dirigent	**2.d. The Inclined Orb**
	2.e. The Maintainer.	
2.c. The Epicycle of the Planet.	**2.f. The Epicycle of the Planet**	**2.e. The Epicycle of the Planet.**
3. The Motions	**3. The Motions**	**3. The Motions**
3.a. The parecliptic	**3.a. The parecliptic**	**3.a. The parecliptic**
3.b. The Eccentric (<u>sequential</u>, i.e., in the <u>direction</u> of the progression of the zodiac).	**3.b. The Eccentric (<u>sequential</u>, i.e., in the <u>direction</u> of the progression of the zodiac).**	**3.b. The Eccentric (<u>sequential</u>, i.e., in the <u>direction</u> of the progression of the zodiac)**
	3.c. The Encompasser: equal to the motion of the Eccentric in magnitude, but opposite in direction, i.e., <u>countersequential</u>.	**3.c. The Encompasser: equal to the motion of the Eccentric in magnitude and direction, i.e., <u>sequential</u>.**

(continued)

(continued)

نهاية الادراك في دراية الافلاك	اختيارات مظفرى	التحفة الشاهية
The Nihāya	*The Ikhtīyārāt*	*The Tuḥfa*
	3.d. The Dirigent	**3.d. The Motion of the Inclined Orb.**
	3.e. The Maintainer.	
3.c. The Epicycle	**3.f. The Epicycle**	**3.e. The Epicycle.**
4. The Three Anomalies of motion.	**4. The Three Anomalies of motion.**	**4. The Three Anomalies of motion.**
	5. A discussion of the equant.	
	6. The solutions involve:	
	6. a. The Ṭūsī couple	
6. b. The Conjectural Hypothesis (based on Appolonius's Theorem)		
6. c. The Hypothesis based on 'Urḍī's Lemma.	**6. c. Hypothesis based on 'Urḍī's Lemma.**	
	7. A discussion involving the Conjectural Hypothesis and the Deductive Hypothesis (based on the 'Urḍī's Lemma, see 6.c) each with two "initial" positions for the center of the epicycle.	
8. A discussion of the merits of the Conjectural Hypothesis	**8. A discussion of the merits of the Conjectural Hypothesis**	
9. An illustration of the Orbs in 2 dimensions and a glossary.	**9. An illustration of the Orbs in 2 dimensions and a glossary.**	**9. An illustration of the Orbs in 2 dimensions and a glossary.**

Appendix C: From the Chapter on the Superior Planets

التذكرة في علم الهيئة *The Tadhkira,* **(from Ṭūsī, *al-Tadhkira fī ʻilm al-hay'a*, vol. 1, 179)**	نهاية الادراك في دراية الافلاك *The Nihāya*	إختيارات مظفرى *The Ikhtīyārāt*
الفصل التاسع فيافلاك الكواكب الباقية و حركاتها الطولية	الباب الثامن في افلاك الكواكب العلوية و الزهرة و حركاتها الطولية	باب هفتم در افلاك كواكب علوى
	لما	
وجدوا الكواكب الثلاثة العلوية أبطأ سيرا من الشمس.	وجدوا الكواكب الثلثة العلوية ابطأ سيرا من الشمس	چون در احوال كواكب علوى تامل كردند و ايشان را بطئ السير تر از آفتاب يافتند
	بحيث	بر وجهى كه
فإذا قارنتها الشمس	اذا قارنتها الشمس	چون آفتاب مقارن ايشان مى شد
		ايشان در بعدى ابعد مى بودند و مستقيم و سريع السير و با سرعت سير ايشان
سبقتها فظهرت مشرقة	سبقتها فظهرت مشرقة	آفتاب در پيش ايشان مى افتاد و ايشان پيش از طلوع آفتاب در جهت مشرق ظاهر مى گشتند
و تكون في اسرع سيرها	و تكون في اسرع سيرها	
ثم تأخذ	ثم تأخذ	
	بَعد التوسط في الحركة	آنگاه سيرشان متوسط مى شد

فيالبطء حتى اذا صارت الشمس إلى قريب من تثليثها الاول أو بعده بقليل وقفت ثم رجعت و تقابلها الشمس فياواسط رجوعاتها ثم تقف ثانيا بقرب وصول الشمس إلى تثليثها الثاني	في البطؤ حتى اذا صارت الشمس الى قريب من تثليثها الاول او بعده بقليل وقفت. ثم رجعت و تقابلها الشمس في اواسط رجوعاتها ثم تقف ثانيا بقُرب وصول الشمس الى تثليثها الثاني	باز بطئ تا چون آفتاب نزديک تثليث اول ايشان می رسيد يا بعد از آن باندکی واقف می شدند آنگاه راجع و در اواسط رجوعات مقابل آفتاب می بودند و در بعدی اقرب باز دوم بار واقف می گشتند بنزديکی رسيدن آفتاب بتثليث دوم ايشان يا
او بعده بقليل	او بعده بقليل	پيش از آن باندکی
ثم تستقيم و تأخذ من البطء	ثم تستقيم و تأخذ من البطؤ	باز مستقيم می گشتند
	الى التوسط	
إلى السرعة الى ان تقرب الشمس منهافتخفتيمغربة	ثم الى السرعة الى ان تقرب الشمسُ منها فتخفي مغربة	
و تقارنها الشمس في اواسط استقاماتها	و تقارنهاالشمس في اوساط استقاماتها	و باز مقارن آفتاب می شدند در اواسط استقامات
	حكموا بان لكل منها تدويرا	حکم کردند که هر يک را فلک تدويريست
	لاستحالة وجود هذه الاحوال بدونه	
	لان مقارناتها للشمس في الابعاد البعيدة و مقابلاتها لها في القريبة و ان امكن	چه مقارنات ايشان با آفتاب در ابعاد بعيده و مقابلات در قريبه اگر چه

	نفرض خارجين او خارج و موافق متحركين الى التوالي حركتين مجموعهما مساو لنصف وسط الشمس	ممکن است که به دو فلک خارج باشد یا خارجی و موافقی که مجموع حرکت ایشان بر توالی مساوی نصف وسط آفتاب بود
	او بخارج فقط حركته ما ذكرنا	یا بیک خارج که حرکت او همین مقدار باشد
	لكن الرجوع و ما يتبعه و كونه في المقارنة في اسرع سيرها يبطل. و الرجوع و ما يتبعه	لکن رجوع و توابع آن و آنک در مقارنه در اسرع سیر می بودندی دافع این احتمال بود
	و ان امكن ان يكون بخارج و موافق	و رجوع و توابع آن اگر چه ممکن است که بخارجی و موافقی بودی
		چنانک در فصل دوم از باب پنجم بیان آن کرده شد لکن اختلاف غایت تعدیل ایشان چه برین تقدیر بسبب آنک بحسب آن بودی که ما بین المرکزین اقتضا کردی و آن یک مقدار معین است متساوی بودی
	لكن كون مقارناتها معها في الابعاد البعيدة و مقابلاتها لها في القريبة يدفعه	و آنک مقارنات ایشان با آفتاب در ابعاد بعیده است و مقابلات او در قریبه مبطل این امکان بود
	و من هذا يعلم ان الاستدلال باحدهما على وجود التدوير	ازین تقریر باید که معلوم کنی که استدلال بیکی ازین دو حکم اعنی مقارنة و مقابلة بر و جه مذکور
	على ما استدل به بعضهم باطل	چنانکه بعضی استدلال کرده اند باطل است
		چه هر یکی بی تدویر صورت می بندد. اما مجموع بی او محالست

	و اما امكان حصول الرجوع و ما يتبعه بدون التدوير و ان سبقت الاشارة اليه في الاصل الرابع لكنا نذكره هاهنا بنمط آخر يشتمل على فوائد خلا عنها المذكور ثمة فنقول متى كانت حركتا الخارج و الموافق مختلفتي الجهة و كانت التي هي الى التوالي اعظم قدرا ليتم للكوكب دورته في البروج و مقهورةً عن التي الى خلاف التوالي في ابعد البعد و قاهرةً اياها في اقربه. و انما يتهيأ هذا بكثرة خروج المركز يلزم منها المطلوب اما الرجوع فحينما تكون زاوية الحركة المرئية الى خلاف التوالي اعظم من المرئية الى التوالي. و اما الاستقامة فعلى العكس و هو ان تكون ٢٠٧المرئية الى التوالي اعظم من المرئية الى خلافه. ثم المرئية العظمى الى التوالي اما ان تكون اصغر من الزاوية الوسطية فيكون مستقيما بطيئا و ان كانت اعظم منها كان مستقيما سريعا و اما الوقوف فعند تكافؤ الحركتين لتكون المرئية الى التوالي كهى الى خلافه. لكن يدفعهما بعد ما مر لزوم كون جرم الكوكب فيالرجوع لكونه في الاوج اصغر منه في الاستقامة لكونها في جانب	و بدانک ابعادی که کواکب را بود از آفتاب که چون بانجا رسند واقف شوند للرجوع يا للاستقامة ایشان را رباطات خوانند و آن درین سه کوکب نزدیکترست بتثلیث دور چنانك گفتیم و در زهره و عطارد بحسب آنک نصف قطر تدویر ایشان اقتضا کند تقریبا چنانك بعد ازین بیاید ان شاء الله

	الحضيض و كون زمان البطؤ و الرجوع لكونه مقدار ما يقطع الكوكب ما بين البعدين الاوسطين بحسب الحركة من جانب الاوج اكثر من زمان السرعة و الاستقامة لكونه مقدار ما يقطع بينهما من جانب الحضيض و كون ما بين اسرع السير و اوسطه اقل من زمان ما بين اوسطه و اقله و كون مقابلاتها مع الشمس في ابعد البعد لانها تقابلها راجعة و كون غايتي تعديلها متساويتين لتعين موضعهما و هو البعد الاوسط بحسب الحركة و الوجود بخلاف الكل هذا ان كانت الحركة التي الي التوالي مقهورة عن التي الي خلافه في ابعد البعد و قاهرة اياها في اقربه و ان فرضت بحيث يكون مقهورية التي الى التوالي مقهورة عن التي الى خلافه في اقرب القرب و قاهريتها ايا ها في ابعده فلا يرد عليه بعد ما مر الاخير	
	و اذ قد ثبت التدوير و عُلم	و چون تدوير ثابت شد و معلوم است
	ان له حاملا و الا لما تكملت دورة الكوكب في البروج	كه اولا حاملى باشد و الا كوكب تمام دور بروج را قطع نكردى
		چنانك در فصل اول از باب پنجم تقرير كرده شد
	عُلِم انه خارج المركز	معلوم شد كه حامل او خارج مركز است
		بچند وجه
	باختلاف غايتي التعديل	يكى آنك باختلاف غايت تعديل

	و اختلاف زمان اختفاء كل منها تحت الشعاع في اجزاء باعيانها من البروج مع ان الاختفاء يكون في الذرى فلا يلحقه من جهة التدوير اختلاف بل من جهة الخارج فيقل زماني الاختفاء عند بعده من الارض و يكثر عند قربه منها لان الشمس هى التي تسبق مركز التدوير فاذا كان ابعد عن الارض كانت حركته ابطأ فتسبقه الشمس اسرع فيقل زمان الاختفاء و يعظم اذا كان اقرب	دوم باختلاف زمان اختفاء هر يك از ايشان در تحت الشعاع در اجزائى معين از فلك البروج با آنك اختفا در ذروة مى باشد پس از جهت تدوير اختلافى لا حق او نشود بل از جهت خارج چون از زمين دور باشد زمان اختفاء اندك شود و چون نزديك باشد بسيار شود چه آفتاب است كه سبق مى گيرد بر مركز تدوير پس چون دور باشد از زمين سير او ابطا بود و سبق اسرع و زمان اختفاء اقل و اگر نزدیک باشد بعكس بود
		سوم
و اذا قيست حال من احواله الى نظير تلك الحال	و باختلاف ايّ حال من احوالها اذا قيست الى نظيرتلك الحال	اختلاف هر حالى از احوال ايشان چون قياس مى كنند با نظير آن حال
	كرجوع الي رجوع او استقامة الى استقامة او بطؤ الي بطؤ او سرعة الى سرعة فانها لا توجد متشابهة بل يكون فيبعض اجزاء البروج اكثر قدرا و زمانا و في بعضها اقل قدرا و زمانا كما في المريخ	چون رجوعى با رجوعى يا استقامتى با استقامتى يا بطئى با بطئى يا سرعتى با سرعتى چه مشابه نمى يابند بل در بعضى اجزاء بروج قدر و زمانش بيشتر مى باشد و در بعضى قدر و زمان كمتر چنانک در مريخ
	و الزهرة	
	او اكبر قدرا و اقل زمانا و بالعكس كما في زحل و المشتري	يا در بعضى قدر بيشتر مى باشد و زمان كمتر و در بعضى بر عكس چنانك در زحل و مشترى
		اين معنى از جداولى كه بجهت قوس رجوع و استقامت و ايام ايشان وضع كرده اند ظاهرست
مخالفة لها		

و الاحوال المتشابهة فيإجزاء بأعيانها من فلك البروج تنتقل بانتقال الثوابت.	ثم حكم لكون الاحوال المتشابهة فياجزاء باعيانها من فلك البروج منتقلة بانتقال الثوابت	باز چون احوال مشابه در اجزاء معين از فلك البروج منتقل يافتند بانتقال ثوابت حكم كردند
	بان الاوج يتحرک بحركة الثوابت	كه خارج (!)بحركت ثوابت منتقل است
و وجدت الاحوال التى يقتضيها البعد الاقرب فياجزاء مقابلة للتي يقتضي فيها البعد الابعد اضدادها		
	و لكون مقتضيات البعد الاقرب في اجزاء مقاطرة التي فيها مقتضيات البعد الابعد بان ابعادها القريبة مقابلة لابعادها البعيدة	و بجهت آنك مقتضيات بعد اقرب در اجزايى يافتند مقاطر آنك در آنجا بود مقتضيات بعد ابعد حكم كردند كه ابعاد قريبه ايشان مقابل ابعاد بعيده است
	و لكونها شمالية عن مدار الشمس في ستة بروج متقاربه اليه تارة و متباعده عنه اخرى و جنوبيه في النصف الآخر كذلك بانّ لها عرضا	و بجهت آنك شمالى بودند از مدار آفتاب در شش برج و جنوبى در باقى و متقارب تارة و متباعد ازو ديگر بار حكم كردند كه ايشانرا عرضى هست
و هيلا تسير على مدار الشمس بعينه بل تكون شماليةعنه فينصف فلك البروج متقاربة إليه تارة و متباعدة عنه أخرى و جنوبية عنه في النصف الآخر كذلك		
	و لكون المجازين منتقلين انتقال الثوابت ايضا بان ممثلاتها متحركة حركة الثوابت	و بسبب انتقال مجازين بانتقال ثوابت حكم كردند كه ممثلات ايشان متحرک است بحركت ثوابت
و المجازان ينتقلان انتقال الثوابت		
	و اعلم ان الرباطات هى ابعاد من	

	الشمس اذا انتهت الكواكب الى حدودها وقفت اما للرجوع او للاستقامة	
	و هو في العلوية قريب من ثلث دائرة كما قلنا و فيالسفليين قريب من نصف قطر تدوير هما كما سيجيء ان شاء الله تعالى	
	ثم وجدوا الزهرة متحركة في الطول لاعلى نفس منطقة البروج بل حواليها تقرب منها تارة في شمالها و تارة في جنوبها و تبعد كذلك لا الى حدين بعينهما فعلم ان لها عرضا مختلفا و ميلا غير ثابت	
وجدوا الزهرة شبيهة الاحوال بعطارد طولا و عرضا الا ان اقرب ابعادها مقابل الا ان اقرب ابعادها مقابل لأبعدها كما في العلوية.		
	ثم وجدوها تُسرع في سيرها فتسبق الشمس بعد مقارنتها و تظهر مغربة ثم بعدالتوسط تاخذ في البطؤ متدرجا الي ان تقف ثم ترجع و تخفى و تقارن الشمس و تفارقها فتسبقها الشمس و تظهر مشرقة ثم تقف و تسقيم من بطؤ الي توسطٍ ثم سرعة الى ان تخفي فتدرك الشمس و تقارنها فتكون معها في منتصفي زماني استقامتها و رجوعها	
وغاية بعدها فيالطول عن الشمس قداما و خلفا لا تتجاوز سبعا و اربعين درجة	و لا تبعد في الطول عنها من قدامها و خلفها اكثر من سبع و اربعين درجة	
	علي الجليل من النظر	
	فحدس انها محمولة على فلك تدوير و ان حركة مركزه موافقة لوسط الشمس و ان	

	التقدم و التخلف لحركة التدوير و انما حكم بان حامل التدوير خارج لما مر و لاختلاف مجموع البعد الصباحي و هو نصف قطر التدوير المار بالبعد الاوسط الذي في نصف الصاعد فيه لظهورها على طرفه قبل طلوع الشمس مشرقة و المسائى و هو النصف الآخر من القطر المذكور لظهورها علي طرفه في اول الليل مغربة و مركز التدوير فيموضع معين من البروج لمجموعهما و مركز التدوير فيموضع آخر منه لدلالته على تقارب مركزالتدوير من مركز العالم و تباعده عنه و الا لما اوتر مجموع البعدين زواياء مختلفة عند مركز العالم و لما اختلف زمان اختفائها ايضا لكنه تختلف فان مركز تدويرها ان كان مسرعا وهى مستقيمة فيقل زمان الاختفاء و ان كانت راجعة فيعظم و ان كان المركز مبطئا و هي مستقيمة فيعظم زمان الاختفاء و ان كانت راجعة فيقل و انما لم يتعرضوا لهذا التفصيل لان عند الاكثرين ان مركز تدويرها لا يسرع و لا يبطئ الا ان تسرع الشمس او تبطئ لتوهمهم ان مركز تدويرها مقارن ابدا لمركز الشمس بالحقيقة و ليس كذلك بل هو بالتقريب و الا لما اختلفت غايتا بعدها الصباحي و المسائي و مركز التدوير في موضع معين نعم قد يقارنه و لهذا قد لاتختلف الغايتان في بعض المواضع و اذا كان كذلك فيتقدم مركز تدويرها على وسط الشمس اذا كان مسرعا و يتاخر عنه ان كان مبطئا و يظهر ما ذكرنا من التفصيل ثم اختلاف اوجيهما و مقدار خروج	

فأثبتوا لكل من الاربعة ثلاثة أفلاك و ثلاث حركات	مركزيهما منع من مقارنة الوسطين ابدا لان رفع الخلاف شرط الوفاق و احوال عطارد في الطول و العرض شبيهة باحوال الزهرة على الوجه المذكور بعينه الا ان اقرب ابعاده لا يقابل ابعدها كما في الزهرة فانه يقابله كالعلوية و كذا حكم الزهرة حكم العلوية في حركة الاوج و انتقال المجازين و لذلك اثبتوا لكل من الاربعة ثلثة افلاك و ثلاث حركات	پس بحسب اين احوال هر يكى را ازين سه سه فلك اثبات كردند لكن چون از احوالى كه هم برصد معلوم كرده بودند چنانک تقرير آن بجاى خويش بيايد تشابه حركت مركز تدوير بود بنسبت با مركز معدل المسير و محاذات ذروه وسطى هم با او و ميل اقطارى كه بذروة حضيض ايشان گذشته باشند از مايل بر وجهى مخصوص و اين احوال از سه فلك حاصل نمى شود لاجرم ما در هر يک ازين كوكب {كواكب} سه فلك ديگر زيادت كرديم تا مجموع شش فلك و شش حركت شد و اين امور بر وجه مراد از انتظام اين افلاك و تركب اين حركات حاصل

Appendix D: The Hypotheses

Anomaly	*Nihāya*	*Tuḥfa*	*Ikhtīyārāt*
I. Fast, intermediate and slow speeds.	1. eccentric orb, اصل الخارج	eccentric orb, اصل الخارج	eccentric orb (Chapter 2.5)
"	2. epicycle, اصل التدوير	epicycle, اصل التدوير	epicycle (Chapter 2.5)
II. Retrograde motion	3. epicycle and deferent اصل التدوير و الحامل	epicycle and deferent اصل التدوير و الحامل	epicycle and deferent (Chapter 2.5)
"	4. An eccentric with a concentric deferent اصل الحامل و الخارج	An eccentric with a concentric deferent اصل الحامل و الخارج	An eccentric with a concentric deferent (Chapter 2.5)
III. Motion uniform about a point other than the center of the mover.	5. The hypothesis of the encompasser: an additional epicycle to make the motion of the new epicycle center uniform about a point other than the center of the original deferent, based on Apollonius's Theorem. اصل المحيطة	Not used	Conjectural Hypothesis اصل حدسی (in Chapter 2.7: The Moon)
"	6. The hypothesis of the maintainer and the dirigent, based on 'Urḍī's Lemma. اصل الحافظة و المدير	اصل المحيطة (not named)	Hypothesis of the Dirigent اصل مدير (in Chapter 2.6: The Sun, and 2.8: The superior planets) Also a species of Deductive Hypothesis اصل استنباطی
IV. Motion of the planet being uniform about a point from which the planet maintained a variable distance	"Based on one of the four also" i.e., an al-'Urḍī or Apollonius configuration. Un-numbered.	Not counted as an anomaly.	One of the four variations included in discussion of an 'Urḍī or Apollonius configuration (in Chapter 2.8: The superior planets)
"	7. The Ṭūsī couple. اصل الصغيرة و الكبيرة	"	The Ṭūsī couple اصل صغيرة و كبيرة (in Chapter 2.7: The Moon)
V. Lack of alignment of the diameter of the planet, because of its motion, with the center of the orb (i.e., the alignment of the epicycle center with the point about which the motion was uniform).	8. The maintainer and the encompasser. الحافظة و المحيطة	Not counted as an anomaly.	(not used as a hypothesis) • درين اصل بكره حافظة محتاج باشيم (In Chapter 2.7: The Moon)
VI. Non-completion of a revolution in the heavens, either in latitude or longitude. (This anomaly is counted as the IVth anomaly for the *Tuḥfa*).	9. "Spherical" Ṭūsī Couple اصل المميل	"Spherical" Ṭūsī Couple اصل المميل	"Spherical" Ṭūsī Couple اصل المميل (In Chapter 2.9: Venus and Mercury)

Appendix E: A Comparison of the *Ikhtīyārāt* and the *Nihāya* in Regard to the Inclination of the Orbs of the Encompasser, the Dirigent, and the Maintainer for the Superior Planets. The *Ikhtīyārāt* Retains an Inclination Scheme that Was Subsequently Revised in the *Nihāya.* The Underlined Text is Crossed-Out in the *Nihāya*.

From the Ikhtīyārāt, MS Aya Sofya 2575, 106v.:	*From the Nihāya*: MS Köprülü 956, 77v.:
	و اما افلاك الكواكب العلوية
	فكل واحد منها يشتمل على ست اكر
	ثلث منها هى الممثل و الحامل و التدوير كما هو عند الجمهور في الحركة و قدرها و جهتها و في وضعها الا في التدوير و الباقية هي التى زدناها
و سيم كره محيطه مركز او بر منطقه حامل و محدب او مماس محدب و مقعر حامل بدو نقطه	اولها الكرة المحيطة و هي في ثخن الحامل بحيث يكون مركزها على منطقتها و يماس سطحها سطحيه على نقطتين
و منطقه او مقاطع منطقه حامل و مايل ازو بقدر غايت ميل ذروه آن كوكب از مايل ميلى ثابت	] ويقاطع منطقتها منطقة الحامل و يميل عنها بقدر غاية ميل ذروة ذلك الكوكب عن المائل ميلا ثابتا [
	ومنقطتها في سطح منطقة الحامل ابدا
و چهارم كره مديره بر مركز محيطه و در اندرون او	و ثانيها الكرة المديرة في جوف المحيطة و على مركزها
و لكن منطقه او در سطح منطقه حامل ابدا	] لكن منطقتها في سطح منطقة الحامل ابدا [

	و محور با محور بر مرکز متقاطع		تقاطع منطقة الحامل وتميل عنها بقدر مثل ذلك الكوكب عن المائل ميلا ثابتا و محورها مقاطع لمحور المحيطة على المركز
	و پنجم کره حافظه در اندرون محیطه بر وجهی که منطقه او که در سطح منطقه محیطه باشد و مرکز از مرکز او خارج بقدر ما بین المرکزین آن کوکب اعنی مرکز عالم و خارج و لکن بشرط آنك این مرکز نقطه بر سطح منطقه محیطه و محور او موازی محور محیطه		و ثالثها الكرة الحافظة في جوف المديرة بحيث يكون منطقتها في سطح منطقة المديرة و مركز ها خارج عن مركز المديرة بقدر خروج مركز حامل ذلك الكوكب عن مرز العالم على ان يكون هذا المركز نقطة على سطح منطقة المديرة و محورها مواز لمحور المديرة
	و منطقه او در سطح منطقه محیطه نه دائما بل وقتی که میل ذروه از مایل در غایت باشد		] و منطقتها في سطح منقطة المحيطة لا دائما بل اذا كان مثل الذروة عن المائل في الغاية [
	چنانك تقریر آن در باب عروض بیاید		

Bibliography

Abū al-Fidā' Ismā'īl ibn 'Alī. 1968. *al-Mukhtaṣar fī akhbār al-bashar*, 1st ed. Baghdād: Maktabat al-Muthanná.

Alhazen. See Ibn al-Haytham.

Allouche, Adel. 1990. Teguder's Ultimatum to Qalawun. *International Journal of Middle East Studies* 22(4): 437–446.

Amitai, R. 1999. Sufis and Shamans: Some remarks on the Islamization of the Mongols in the Ilkhanate. *Journal of the Economic and Social History of the Orient* 42(1): 27–46.

Amitai, R. 2000. "ḠĀZĀN KHAN, MAḤMŪD." *Encyclopaedia Iranica*. Encyclopaedia Iranica Online, December 15, 2000. http://www.iranica.com/articles/gazan-khan-mahmud.

Amitai, R. 2004. "HULĀGU KHAN." *Encyclopaedia Iranica*. Encyclopaedia Iranica Online, December 15, 2004. http://www.iranica.com/articles/hulagu-khan.

Aqsārā'ī, Karīm al-Dīn. 1983. Tārīkh-*i Salājiqah, yā, Musāmarat al-akhbār wa musāyarāt al-akhyār*. Tehran: Intisharat-i Asatir.

Aristotle. 1956. *The metaphysics*. Cambridge, MA: Harvard University Press.

Banākatī, Dāwūd ibn Muḥammad. 1348. *Tārīkh-i Banākatī = Rawḍat ūlā al-albāb fī ma'rifat al-tawārīkh wa al-ansāb*. Tehrān.

Bar Hebraeus. 1890. *Tārīkh mukhtaṣar al-duwal*. Bayrūt: al-Maṭba'ah al-Kāthūlīkīyah lil-Ābā' al-Yasū'īyīn.

Bar Hebraeus. 1932. *The Chronography of Gregory Abû'l Faraj, the Son of Aaron, the Hebrew Physician, Commonly Known as Bar Hebraeus*, ed. E. A. Wallis Budge. London: Oxford University Press, H. Milford.

Barthold, W. 2011. "DJuwaynī, 'Alā' al-Dīn 'Aṭā-Malik b. Muḥammad." *Encyclopaedia of Islam*, 2nd ed. Brill Online, 2011. http://www.brillonline.nl/subscriber/entry?entry=islam_SIM-2131.

Biran, Michal. 2009. "JOVAYNI, ṢĀḤEB DIVĀN." *Encyclopaedia Iranica*. June 15, 2009. http://www.iranica.com/articles/jovayni-saheb-divan.

Bīrūnī, Muḥammad ibn Aḥmad. 1362. *Kitāb al-tafhīm li-awā'il ṣinā'at al-tanjīm*. Tihrān: Intishārāt-i Bābak.

Bīrūnī, Muḥammad ibn Aḥmad. 1878. *Chronology of ancient nations; An English version of the Arabic text of the Athâr-ul-Bâkiya of Albîrûnî, or "Vestiges of the past"*. Leipzig: F. A. Brockhaus.

Bīrūnī, Muḥammad ibn Aḥmad. 1910. *Alberuni's India. An account of the religion, philosophy, literature, geography, chronology, astronomy, customs, laws and astrology of India about A.D. 1030*. An English ed. London: K. Paul, Trench, Trübner & Co., ltd.

Bīrūnī, Muḥammad ibn Aḥmad. 1958. *Kitāb fī taḥqīq mā lil-Hind min maqbūlah maqbūlah fī al-'aql aw mardhūlah*. Hyderabad: Osmania Oriental Publications Bureau.

K. Niazi, *Quṭb al-Dīn Shīrāzī and the Configuration of the Heavens: A Comparison of Texts and Models*, Archimedes 35, DOI 10.1007/978-94-007-6999-1,

Bīrūnī, Muḥammad ibn Aḥmad. 1973. *Āl-Biruniś book on pharmacy and materia medica*. Karachi: Hamdard Academy.

Bīrūnī, Muḥammad ibn Aḥmad. 1993. *Kitāb fī taḥqīq mā li al-Hind*. Vol. 105. Publications of the Institute for the History of Arabic Islamic Science. Frankfurt am Main: Johann Wolfgang Goethe University.

Blair, Sheila S. 1986. The Mongol capital of Sulṭānīya, 'The Imperial'. *Iran* 24: 139–151.

Boilot, D. 2011. "al-Bīrūnī (Bērūnī) Abu 'l-Rayḥān Muḥammad b. Aḥmad." *Encyclopaedia of Islam*, 2nd ed. Brill Online, 2011. http://www.brillonline.nl/subscriber/entry?entry=islam_SIM-1438.

Bosworth, C. 2010a. "Ordu." *Encyclopaedia of Islam*, 2nd ed. Brill Online, 2010. http://www.brillonline.nl/subscriber/entry?entry=islam_COM-0879.

Bosworth, C. 2010b. "Salghurids." *Encyclopaedia of Islam*, 2nd ed. Brill Online, 2010. http://www.brillonline.nl/subscriber/entry?entry=islam_SIM-6531.

Boyle, John Andrew. 1968. Dynastic and political history of the Īl-khāns. In *The Cambridge history of Iran*, vol. 5, ed. J.A. Boyle, 303–421. Cambridge: Cambridge University Press.

Brockelmann, Carl. 1937. *Geschichte Der Arabischen Litteratur von Prof. Dr. C. Brockelmann: Erster [—Dritter] Supplementband*. Leiden: Brill.

Bulliet, Richard. 2009. *Cotton, climate, and camels in early Islamic Iran*. New York: Columbia University Press.

Bulliet, Richard. 2011. Abu Muslim and Charlemagne. In *Community, state, history and changes: Festschrift for Prof. Ridwan al-Sayyid*, 19–28. Beirut: Arab Network for Research and Publishing.

Cahen, Claude. 1968. *Pre-Ottoman Turkey a general survey of the material and spiritual culture and history C. 1071–1330*. New York: Taplinger Pub. Co.

Cooper, J. C. (Jean C.). 1984. *Chinese alchemy: The Taoist quest for immortality*. Wellingborough, Northamptonshire: Aquarian Press.

Daftary, Farhad. 1990. *The Ismāʿīlīs: Their history and doctrines*. Cambridge: Cambridge University Press.

Dhahabī, Muḥammad ibn Aḥmad. 1987. *Tārīkh al-Islām wa-wafayāt al-mashāhīr wa al-aʿlām*, ed. U. Tadmurī. Bayrūt: Dār al-Kitāb al-ʿArabī.

Encyclopaedic Ethnography of Middle-East and Central Asia. 2005. 1st ed. New Delhi: Global Vision Publishing House.

Evans, James. 1998. *The history & practice of ancient astronomy*. New York: Oxford University Press.

Fakhr al-Dīn Gurgānī. 1865. *Masnavi-i Vis va Ramin*. Calcutta: College Press.

Fleisch, H. 2010. "Ibn al- Ḥādjib , Djamāl al-Dīn Abū 'Amr 'Uthmān b. 'Umar b. Abī Bakr al-Mālikī." *Encyclopaedia of Islam*, 2nd ed. Brill Online, 2010. http://www.brillonline.nl/subscriber/entry?entry=islam_COM-0324.

Gamini, Amir Mohammad. 1388. The planetary models of Quṭb al-Dīn Shīrāzī in the Ikhtīyārāt-i Muẓaffarī. *Tarīkh-i ʿilm* 8: 39–54.

Gamini, Amir Mohammad, and H. Masoumi Hamedani. 2013. Al-Shīrāzī and the empirical origin of Ptolemy's Equant in his model of the superior planets. *Arabic Sciences and Philosophy* 23(1): 47–67.

Goldstein, Bernard R. 1967. The Arabic version of Ptolemy's planetary hypotheses. *Transactions of the American Philosophical Society* 57(4): 3–55. New Series.

González de Clavijo, Ruy. 2001. *Narrative of the embassy of Ruy Gonzalez De Clavijo to the Court of Timour at Samarcand, A.D. 1403–6: Translated for the first time with notes, a Preface, and an Introductory Life of Timour Beg*. New Delhi: Asian Educational Services.

Ḥāfiẓ Abrū. 1349. *Jughrāfīyā-yi Ḥāfiẓ Abrū: Qismat-i Rub-ʿi Khurāsān, Harāt*, ed. R. Mayil Haravii. Tehran: Bunyād-i Farhang-i Īrān.

Hairi, Abdul-Hadi. 2010. "Khwānsārī , Sayyid Mīrzā Muḥammad Bāḳir Mūsawī Čahārsūḳī b. Mīrzā Zayn al-ʿĀbidīn." *Encyclopaedia of Islam*, 2nd ed. Brill Online, 2010. http://www.brillonline.nl/subscriber/entry?entry=islam_SIM-4193.

Ḥamd Allāh Mustaufī Qazvīnī. 1915. *The Geographical Part of the Nuzhat al-Qulūb Composed by Ḥamd-Allāh Mustawfī of Qazwīn in 740 (1340)*, ed. G. Le Strange. Leyden: E.J. Brill.
Ḥamd Allāh Mustaufī Qazvīnī. 1960. *Tārīkh-i Guzīdah*, ed. A. Nawa'i. Tehran: Amir Kabir.
Hartner, Willy. 1950. The Astronomical instruments of Cha-ma-lu-ting, their identification, and their relations to the instruments of the observatory of Marāgha‡. *Isis* 41(2): 184–194.
Hodgson, Marshall G.S. 1974. *The venture of Islam: Conscience and history in a world civilization*. Chicago: University of Chicago Press.
Holt, P.M. 1986. The Īlkhān Aḥmad's embassies to Qalāwūn: Two contemporary accounts. *Bulletin of the School of Oriental and African Studies, University of London* 49(1): 128–132.
Ibn 'Abd al-Ẓāhir, Muḥyī al-Dīn. 1961. *Tashrīf al-ayyām wa al-'uṣūr fī sīrat al-malik al-manṣūr*, 1st ed. al-Qāhirah: Wizārat al-thaqāfah wa-al-irshād al-qawmī, al-idārah al-'āmmah lil-thaqāfah.
Ibn al-Athīr, 'Izz al-Dīn. 1385. *al-Kāmil fī al-tārīkh*. Báyrūt: Dar Ṣáder.
Ibn al-Fuwaṭi, 'Abd al-Razzaq ibn Ahmad. 1995. *Majma' al-adab fi mu'jam al-alqab*, ed. M. Kazim. Tehran: Muassasat al-tiba' ah wa al-nashr, Wizarat al- thaqafah wa al-irshad al-Islami.
Ibn al-Haytham, Abū ʿAlī al-Ḥasan. 1971. *Shukūk 'alá Baṭlamyūs*. al-Qāhirah: Maṭba'at Dār al-Kutub.
Ibn al-Haytham, Abū ʿAlī al-Ḥasan. 1990. *Ibn al-Haytham's On the Configuration of the World*, ed. Tzvi Langermann. New York: Garland.
Ibn 'Asākir, Shāfi' ibn 'Alī. 2000. *Šāfi' Ibn Alī's Biography of the Mamluk Sultan Qalāwūn*. Warsaw: Dialog.
Ibn Battutah, Abu Abdallah. 2002. *The Travels of Ibn Battutah*. London: Picador.
Ibn Bībī, Nāṣir al-Dīn Ḥusayn ibn Muḥammad. 1902. *Histoire des Seldjoucides d'Asie mineure d'après l'abrégé du Seldjouknāmeh d'Ibn-Bībī: Texte Persan, Publié d'après le Ms. de Paris*. Leiden: E. J. Brill.
Ibn Bībī, Nāṣir al-Dīn Ḥusayn ibn Muḥammad. 1971. *Akhbār-i Salājiqah-i Rūm*, 1st ed. Tehrān: Kitābfurūshī-i Tehrān.
Ibn Ḥajar al-'Asqalānī, Aḥmad ibn 'Alī. 1966. *al-Durar al-kāminah fī a'yān al-mi'ah al-thāminah*, 2nd ed. al-Qāhirah: Dār al-Kutub al-Ḥadīthah.
Ibn Khaldūn. 1945. *The Muqaddima*. Cairo: Matba'at Mustafa Muhammad.
Iqbāl, 'Abbās. 1365. *Tārīkh-i Mughūl: az ḥamlah-'i Changīz tā tashkīl-i dawlat-i Taymūrī*, 6th ed. Tihrān: Amīr Kabīr.
Isnawī, 'Abd al-Raḥīm ibn al-Ḥasan. 1971. *Ṭabaqāt al-Shāfi'īyah*. Baghdad: Riasat diwan al-awqaf.
Jackson, P. 1982. "ABAQA." *Encyclopaedia Iranica*. December 15, 1982. http://www.iranica.com/articles/abaqa.
Jackson, P. 1984. "AḤMAD TAKŪDĀR." *Encyclopaedia Iranica*. December 15, 1984. http://www.iranica.com/articles/ahmad-takudar-third-il-khan-of-iran-r.
Jackson, P. 1986. "ARḠŪN KHAN." *Encyclopaedia Iranica*. December 15, 1986. http://www.iranica.com/articles/argun-khan-fourth-il-khan-of-iran-r683-90-1284-91.
Juvaynī, 'Alā' al-Dīn 'Aṭā Malik. n.d. *The Ta'rikh-i Jahán-Gushá*, ed. M. Qazvini. Tehran: Bamdad.
Juvaynī, 'Alā' al-Dīn 'Aṭā Malik. 1371. *Jahāngushā-yi Juvaynī: Changīz, Tārābī, Khvārazmshāh, Ḥasan Ṣabbāḥ, bā ma'nī-i vāzhah'hā*, ed. K. Khatib Rahbar. Tehrān: Intishārāt-i Mahtāb.
Juvaynī, 'Alā' al-Dīn 'Aṭā Malik. 1997. *Genghis Khan: The history of the world conqueror*, ed. J. A. Boyle. Seattle: University of Washington Press.
Kāshānī, 'Abd Allāh ibn 'Alī. 1348. *Tārīkh-i Ūljāyatū, Tārīkh-i Pādishāh-i Sa'īd Ghiyāth al-dunyá va al-Dīn Uljāyitū Sulṭān Muḥammad*. Tehrān: Bungāh-i Tarjumah va Nashr-i Kitāb.
Kātip Çelebi. 1992. *Kashf al-ẓunūn 'an asāmī al-kutub wa-al-funūn*. Bayrūt, Lubnān: Dār al-Kutub al-'Ilmīyah.
Kennedy, E.S. 1966. Late medieval Planetary theory. *Isis* 57(3): 365–378.
Kennedy, E.S. 1968. The exact sciences in Iran under the Saljuqs and Mongols. In *Cambridge History of Iran*, 659–679. Cambridge: Cambridge University Press.

Kermani, Afḍal al-Dīn. 1326. *Badāyi' al-zamān fi waqāyi' Kirmān*, ed. M. Bayani. Tehran: Intisharat-i daneshgah-e Tehran.

Khvansari, Muhammad Baqir. 1962. *Rawdāt al-jannāt fi aḥwāl al-'ulamā wa al-sadāt*, ed. M. Rawdati. Tehran: Dar al-Kutub al-Islamiyah.

Kolbas, Judith G. 2006. *The Mongols in Iran: Chingiz Khan to Uljaytu, 1220–1309*. London: Routledge.

Komaroff, Linda. 2002. Introduction: On the Eve of the Mongol Conquest. In *The legacy of Genghis Khan: Courtly art and culture in Western Asia, 1256–1353*, 3–7. New York: Metropolitan Museum of Art.

Komaroff, Linda (ed.). 2006. *Beyond the legacy of Genghis Khan*. Leiden: Brill.

Köprülü, Mehmet Fuat. 1992. *The Seljuks of Anatolia: Their history and culture according to local Muslim sources*. Salt Lake City: University of Utah Press.

Kutubī, Muḥammad ibn Shākir. 1973. *Fawāt al-wafāyāt wa al-dhayl 'alayhā*. Beirut: Dar al-Thaqafah.

Lambton, Ann K.S. 1988. *Continuity and change in medieval Persia: Aspects of administrative, economic, and social history, 11th–14th century*. Albany: Bibliotheca Persica.

Lambton, Ann K.S. 1991. *Landlord and Peasant in Persia: A study of land tenure and land revenue administration*. London: I.B. Tauris.

Lane, George. 2004. *Genghis Khan and Mongol Rule*. Westport: Greenwood Press.

Lane, George. 2009. "JOVAYNI, 'ALĀ'-AL-DIN." *Encyclopaedia Iranica*. June 15, 2009. http://www.iranica.com/articles/jovayni-ala-al-din.

Leclerc, Lucien. 1876. *Histoire de la Médecine Arabe*. Paris: E. Leroux.

Li, Zhichang. 1931. *The travels of an Alchemist; The journey of the Taoist, Ch'ang-Ch'un, from China to the Hindukush at the Summons of Chingiz Khan, recorded by his disciple, Li Chih-Ch'ang*. London: G. Routledge & Sons Ltd.

Livingston, J. 1973. Nasir al-Din al-Tusi's al-Tadhkirah. *Centaurus* 17: 260–275.

Lockhart, L. 2010. "DJibāl." *Encyclopaedia of Islam*, 2nd ed. Brill Online, 2010. http://www.brillonline.nl/subscriber/entry?entry=islam_SIM-2068.

MacDonald, D.B. 2013. "Ḳārūn." *Encyclopaedia of Islam*, 2nd ed. Brill Online, 2013. http://referenceworks.brillonline.com/entries/encyclopaedia-of-islam-2/karun-SIM_3951.

MacKenzie, D. 2010a. "CHORASMIA iii. The Chorasmian Language." *Encyclopaedia Iranica*, 2010. http://www.iranica.com/articles/chorasmia-iii.

MacKenzie, D. 2010b. "Iran iii. Languages." *Encyclopaedia of Islam*, 2nd ed. Brill Online, 2010. http://www.brillonline.nl/subscriber/entry?entry=islam_COM-1409.

Madelung, W. 2010. "al-Zamakhsharī, Abu 'l-Ḳāsim Maḥmūd b. 'Umar (Contributions in the fields of theology, exegesis, ḥadīth and adab)." *Encyclopaedia of Islam*, 2nd ed. Brill Online, 2010. http://www.brillonline.nl/subscriber/entry?entry=islam_COM-1469.

Maqrīzī, Aḥmad ibn 'Alī. 1997. *al-Sulūk li ma'rifat duwal al-mulūk*, 1st ed. Bayrūt: Dār al-Kutub al-'Ilmīyah.

Martin, H. Desmond. 1950. *The rise of Chingis Khan and his conquest of North China*. Baltimore: Johns Hopkins Press.

Melville, C. n.d. The Īlkhān Öljeitü's conquest of Gīlān (1307): Rumour and reality. In *The Mongol empire and its legacy*, ed. Reuven Amitai-Preiss and David O. Morgan. Leiden: Brill.

Melville, C. 2009. Anatolia under the Mongols. In *Byzantium to Turkey, 1071–1453*, The Cambridge history of Turkey, vol. 1, 51–101. Cambridge: Cambridge University Press.

Minorsky, V. 2010. "Sulṭānīya." *Encyclopaedia of Islam*, 2nd ed. Brill Online, 2010. http://www.brillonline.nl/subscriber/entry?entry=islam_COM-1118.

Minorsky, V. 2011a. "Marāgha." *Encyclopaedia of Islam*, 2nd ed. Brill Online, 2011. http://www.brillonline.nl/subscriber/entry?entry=islam_COM-0676.

Minorsky, V. 2011b. "al-Rayy." *Encyclopaedia of Islam*, 2nd ed. Brill Online, 2011. http://www.brillonline.nl/subscriber/entry?entry=islam_COM-0916.

Minovi, M. 1348. Mulla Qutb Shirazi. In *Yadnameh-i Irani-i Minorsky*, 165–205. Tehran: Intisharat-i daneshgah-i Tehran.

Mir, Muhammad Taqi. 1977. *Sharḥ-i ḥal wa āsār-i 'allamah Qutb al-Dīn Maḥmud Ibn Mas'ud Shīrāzī, danishmand-i 'ali qadr-i qarn-i haftum, (634–710 A.H.)*. Intisharat-i Danishgah-i Pahlavi 91. Shiraz: Danishgah-i Pahlavi.

Mohaghegh, M. 2010. "al- Kātibī, Nadjm al- Dīn Abu'l-Ḥasan 'Alī b. 'Umar." *Encyclopaedia of Islam*, 2nd ed. Brill Online, 2010. http://www.brillonline.nl/subscriber/entry?entry=islam_SIM-4023.

Morgan, D. 1988. *Medieval Persia, 1040–1797*. London: Longman.

Morgan, D. 2007. *The Mongols*, 2nd ed. Malden: Blackwell Publishing.

Morgan, D. 2010. "Öldjeytü." *Encyclopaedia of Islam*, 2nd ed. Brill Online, 2010. http://www.brillonline.nl/subscriber/entry?entry=islam_SIM-6018.

Morgan, D. 2011. "Rashīd al-Dīn Ṭabīb." *Encyclopaedia of Islam*, 2nd ed. Brill Online, 2011. http://www.brillonline.nl/subscriber/entry?entry=islam_SIM-6237.

Morrison, Robert. 2005. Qutb al-Din al-Shirazi's hypotheses for celestial motions. *Journal for the History of Arabic Science* 13.

Morrison, Robert. 2007. *Islam and science: The intellectual career of Niẓām al-Dīn al-Nīsāburī*. London: Routledge.

Morton, A.H. n.d. The letters of Rashid al-Din: Ilkhanid fact or Timurid fiction? In *The Mongol empire and its legacy*, ed. R. Amitai-Preiss. Leiden: Brill.

Mudarris Razavi, Muhammad. 1955. *Aḥwāl wa Athār-i Muḥammad Ibn Muḥammad Ibn al-Ḥasan al-Ṭūsī*. Tehran: Intisharat-i daneshgah-i Tehran.

Nasawī, Muḥammad ibn Aḥmad. 1986. *Sirat Jalal Al-Din Minkubirni*, ed. Minovi, Mujtabá. Tihran: Shirkat-i Intisharat-i 'Ilmi va Farhangi, vabastah bih Vizarat-i Farhang va Amuzish-i 'Ali.

Nasr, S.H. 1996. *The Islamic intellectual tradition in Persia*. Richmond: Curzon Press.

Nasr, S.H. 1999a. Life sciences, alchemy and medicine. In *The Cambridge history of Iran*, vol. 4, ed. Richard Frye, 396–418. Cambridge: Cambridge University Press.

Nasr, S.H. 1999b. Philosophy and cosmology. In *The Cambridge history of Iran*, vol. 4, ed. Richard Frye, 419–441. Cambridge: Cambridge University Press.

Needham, Joseph. 1954. *Science and civilisation in China*. Cambridge: Cambridge University Press.

Neugebauer, Otto. 1957. *The exact sciences in antiquity*. Providence: Brown University Press.

Onon, Urgunge. 2001. *The secret history of the Mongols: The life and times of Chinggis Khan*. Richmond, Surrey: Curzon.

Pedersen, Olaf. 1974. *A survey of the Almagest*. Odense: Odense Universitetsforlag.

Petrushevsky, I. 1344. *Kishāvarzī va munāsabāt-i arzī dar Īran-i ahd-i Mughūl, Qarnhā-yi 13 va 14 mīlādī*. Tehran: Mu'assasah-'i muṭāla'āt va taḥqīqāt-i ijtimā'ī.

Petrushevsky, I. 1968. The socio-economic condition of Iran under the Il-Khans. In *Cambridge history of Iran*, vol. 5, ed. J.A. Boyle, 483–537. Cambridge: Cambridge University Press.

Pingree, David. 1998. "E<u>K</u>TĪĀRĀT." *Encyclopaedia Iranica*. December 15, 1998. http://www.iranica.com/articles/ektiarat.

Prestwich, Michael. 1997. *Edward I*. New Haven: Yale University Press.

Ptolemy. 1998. *The Almagest*. Princeton: Princeton University Press.

Ragep, F. Jamil. 2000. The Persian context of the Ṭūsī couple. In *Naṣīr al-Dīn Ṭūsī, Philosophe et Savant du XIIIe Siècle*, 113–130. Teheran: Presses Universitaires D'Iran.

Ragep, F. Jamil. 2004. Ibn al-Haytham and Eudoxus: The revival of homocentric modeling in Islam. In *Studies in the history of the exact sciences in honour of David Pingree*, ed. Burnett Charles, Jan P. Hogendijk, Kim Plofker, and Michio Yano, 786–809. Leiden: E. J. Brill.

Ragep, F. Jamil, and B. Hashemipour. 2006. Juft-i Ṭūsī (the Ṭūsī couple). In *The Encyclopaedia of the World of Islam*, vol. X, 472–475. Tehran: Encyclopaedia Islamica Foundation.

Rashīd al-Dīn Ṭabīb. 1338. *Jāmi' al-tawārīkh*, ed. Bahman Karimi. Tehrān: Iqbāl.

Rashīd al-Dīn Ṭabīb. 1947. *Mukatabat-e Rashidi*. Lahore: Panjab University Oriental Publications.

Rashīd al-Dīn Ṭabīb. 1998. *Rashiduddin Fazlullah's Jami'u't-Tawarikh = Compendium of Chronicles*. Cambridge, MA: Harvard University, Department of Near Eastern Languages and Civilizations.

Robson, J. 2010. "al- Baghawī, Abū Muhammad al-Ḥusayn b. Mas'ūd b. Muḥ. al-Farrā' (or Ibn al-Farrā')." *Encyclopaedia of Islam*, 2nd ed. Brill Online, 2010. http://www.brillonline.nl/subscriber/entry?entry=islam_SIM-1024.

Rosenthal, F. 1968. *A history of Muslim historiography*, 2nd ed. Leiden: E. J. Brill.

Rosenthal, F. 2010a. "Ibn al- Athīr." *Encyclopaedia of Islam*, 2nd ed. Brill Online, 2010. http://www.brillonline.nl/subscriber/entry?entry=islam_SIM-3094.

Rosenthal, F. 2010b. "Ibn al- Fuwaṭī , Kamāl al-Dīn 'Abd al-Razzāḳ b. Aḥmad." *Encyclopaedia of Islam*, 2nd ed. Brill Online, 2010. http://www.brillonline.nl/subscriber/entry?entry=islam_SIM-3165.

Rosenthal, F. 2010c. "Ibn Ḥadjar al- 'Asḳalānī , Shihāb al-Dīn Abu 'l-Faḍl Aḥmad b. Nūr al-Dīn 'Alī b. Muḥammad." *Encyclopaedia of Islam*, 2nd ed. Brill Online, 2010. http://www.brillonline.nl/subscriber/entry?entry=islam_SIM-3178.

Rossabi, Morris. 2002. The Mongols and their legacy. In *The legacy of Genghis Khan: Courtly art and culture in Western Asia, 1256–1353*, 13–35. New York: Metropolitan Museum of Art.

Sadeque, Syedah Fatima. 1956. *Baybars I of Egypt*, 1st ed. Dacca: Oxford University Press.

Şaik Gökyay, Orhan. 2010. "Kātib Čelebi , appellation of muṣṭafā b. 'abd allāh (1017- 67/1609-57), known also (after his post in the bureaucracy) as ḥādjdjī khalīfa." *Encyclopaedia of Islam*, 2nd ed. Brill Online, 2010. http://www.brillonline.nl/subscriber/entry?entry=islam_COM-0467.

Safi, Omid. 2006. *The politics of knowledge in premodern Islam: Negotiating ideology and religious inquiry*. Chapel Hill: University of North Carolin Press.

Saliba, George. 1979. The first non-Ptolemaic astronomy at the Maraghah School. *Isis* 70(4): 571–576.

Saliba, George. 1991. Persian scientists in the Islamic world. In *The Persian presence in the Islamic World*, ed. Richard G. Havannisian and Georges Sabagh, 126–146. Cambridge: Cambridge University Press.

Saliba, George. 1994. *A history of Arabic astronomy: Planetary theories during the golden age of Islam*. New York: New York University Press.

Saliba, George. 1996. Arabic planetary theories after the 11th century AD. In *Encyclopaedia of the history of Arabic Science*, ed. Roshdi Rashed, 58–127. London: Routledge.

Saliba, George. 2004. Aristotelian cosmology and Arabic astronomy. In *De Zénon d'Elée à Poincaré: Recueil d'études en hommage à Roshdi Rashed*, 251–268. Louvain: Peeters.

Saliba, George. 2006. Horoscopes and Planetary theory: Ilkhanid patronage of Astronomers. In *Beyond the legacy of Genghis Khan*, 357–368. Leiden: Brill.

Saliba, George. 2007. *Islamic science and the making of the European renaissance*. Cambridge, MA: MIT Press.

Sallāmī, Muḥammad ibn Rāfi'. 2000. *Tārīkh 'ulamā' Baghdād al-musammá muntakhab al-mukhtār*, 2nd ed. Bayrūt: Dār al-'Arabīyah lil-Mawsū'āt.

Sambursky, Samuel. 1987. *The physical world of late antiquity*. London: Routledge & Kegan Paul.

Samsó, J. 2010. "Marṣad." *Encyclopaedia of Islam*, 2nd ed. Brill Online, 2010. http://www.brillonline.nl/subscriber/entry?entry=islam_SIM-4972.

Sarton, George. 1962. *Introduction to the history of science*. Baltimore: Pub. for the Carnegie Institution of Washington, by the Williams & Wilkins Co.

Sayf ibn Muḥammad ibn Ya'qūb, al-Harawī. 1944. *The Ta'ríkh Náma-I-Harat (The history of Harát) of Sayf Ibn Muḥammad Ibn Ya'qúb Al-Harawí*. Calcutta: Baptist Mission Press.

Sayılı, Aydın. 1960. *The observatory in Islam and its place in the general history of the observatory*. Ankara: Türk Tarih Kurumu Basımevi.

Schact, J. 2010. "al- Subkī." *Encyclopaedia of Islam*, 2nd ed. Brill Online, 2010. http://www.brillonline.nl/subscriber/entry?entry=islam_SIM-7116.

Shawkānī, Muḥammad ibn 'Alī. 1978. *al-Badr al-ṭāli' bi-maḥāsin man ba'da al-qarn al-sābi'*. Bayrūt, Lubnān: Dār al-Ma'rifah.

Shīrāzī, Quṭb al-Dīn. *Ikhtīyārāt-i Muẓaffarī*, Ayasofya MS 2575, Istanbul.

Shīrāzī, Quṭb al-Dīn. *Ikhtīyārāt-i Muẓaffarī*, Majlis MS 6398, Tehran.

Shīrāzī, Quṭb al-Dīn. *Nihāyat al-idrāk fī dirāyat al-aflāk*, Köprülü MS 956, Istanbul.

Shīrāzī, Quṭb al-Dīn. *Nihāyat al-idrāk fī dirāyat al-aflāk*, Köprülü MS 957, Istanbul.

Shīrāzī, Quṭb al-Dīn. *al-Tuḥfa al-sa'dīya fī al-ṭibb,* Suleimaniye MS 3649, Istanbul.
Shīrāzī, Quṭb al-Dīn. *al-Tuḥfa al-shāhiyya,* BN Arabe MS 2516, Paris.
Shīrāzī, Quṭb al-Dīn. *Durrat al-tāj li-ghurrat al-dabāj*, Majlis MS 4729, Tehran.
Shīrāzī, Quṭb al-Dīn. 1320. *Durrat al-tāj li-ghurrat al-dabāj.* Tehran: Chāpkhāneh-i Majlis.
Shīrāzī, Quṭb al-Dīn. 2001. *Sharh-i Hikmat al-Ishraq-i Suhravardi.* Nurani, 'Abd Allah. Silsilah-i danish-i Irani; 50; Tihran: Muassasah-i Mutala'at-i Islami, Danishgah-i Tihran, Danishgah-i Mak'gil.
Shīrāzī, Quṭb al-Dīn. 2003. *Bayan al-hajah ilá al-tibb wa al-atibba wa adabuhum wa wasayahum.* Bayrut: Dar al-Kutub al-'Ilmiyah.
Soudavar, A. 2003. In defense of Rashid-od-Din and his letters. *Studia Iranica* 23: 77–120.
Spuler, B. 2010a. "DJuwaynī, Shams al-Dīn Muḥammad b. Muḥammad." *Encyclopaedia of Islam*, 2nd ed. Brill Online, 2010. http://www.brillonline.nl/subscriber/entry?entry=islam_SIM-2132.
Spuler, B. 2010b. "Gaykhātū." *Encyclopaedia of Islam*, 2nd ed. Brill Online, 2010. http://www.brillonline.nl/subscriber/entry?entry=islam_SIM-2427.
Storey, C. 1958. *Persian literature: A bio-bibliographical survey.* London: Luzac and Co., Ltd.
Subkī, Tāj al-Dīn 'Abd al-Wahhāb ibn 'Alī. 1324. *Ṭabaqāt al-Shāfi'īyah al-Kubrá.* 1st ed. al-Qāhirah: al-Maṭba'ah al-Ḥusaynīyah.
Suter, H. 1900. *Die Mathematiker und Astronomen Der Araber und Ihre Werke.* Leipzig: B.G. Teubner.
Swerdlow, N.M. 1973. The derivation and first draft of Copernicus's Planetary theory: A translation of the Commentariolus with commentary. *Proceedings of the American Philosophical Society* 117(6): 423–512.
Swerdlow, N.M. 1984. *Mathematical astronomy in Copernicus's De Revolutionibus.* New York: Springer.
Tauhid, Ahmad. 1910. "Rum Seljuqu daulatinin inqiraz-ile teshkil eden tawa'if muluk (ma ba'd)." *Tarih "Uthmani Encumeni Mecmu" esi* 5: 317–321.
Turan, O. 1970. Anatolia in the period of the Seljuks and the Beyliks. In *The Cambridge history of Islam*, vol. 1. Cambridge: Cambridge University Press.
Ṭūsī, Naṣīr al-Dīn Muḥammad ibn Muḥammad. *Ḥall-i mushkilāt-i mu'īnīyah*, Majlis MS 6346, Tehran.
Ṭūsī, Naṣīr al-Dīn Muḥammad ibn Muḥammad. 1335a. *Ḥall-i mushkilāt-i mu'īnīyah.* Tehran: Chāpkhānah-'i Dānishgāh.
Ṭūsī, Naṣīr al-Dīn Muḥammad ibn Muḥammad. 1335b. *al-Risālah al-mu'īnīyah.* Tehran: Chāpkhānah-'i Dānishgāh.
Ṭūsī, Naṣīr al-Dīn Muḥammad ibn Muḥammad. 1993. *Naṣīr al-Dīn al-Ṭūsī's Memoir on Astronomy = al-Tadhkira fī 'ilm al-hay'a*, ed. F. Jamil Ragep. New York: Springer.
Ṭūsī, Naṣīr al-Dīn Muḥammad ibn Muḥammad. 1994. *Zubdat al-idrāk fī hay'at al-aflāk: ma'a dirāsah li-manhaj al-Ṭūsī al-'ilmī fī majāl al-falak.* 1st ed. al-Iskandarīyah: Dār al-Ma'rīfah al-Jāmi'īyah.
Urḍī, Mu'ayyad al-Dīn Ibn Burayk, al-. 1990. *Kitāb al-hay'a*, ed. George Saliba. Bayrūt: Markaz Dirāsāt al-Waḥdah al-'Arabīyah.
Vajda, G. 2010. "Idjāza." *Encyclopaedia of Islam*, 2nd ed. Brill Online, 2010. http://www.brillonline.nl/subscriber/entry?entry=islam_SIM-3485.
Versteegh, C. 2010. "al- Zamakhsharī, Abu 'l- Ḳāsim Maḥmūd b. 'Umar." *Encyclopaedia of Islam*, 2nd ed. Brill Online, 2010. http://www.brillonline.nl/subscriber/entry?entry=islam_SIM-8108.
Virani, Shafique N. 2003. The eagle returns: Evidence of continued Ismaili activity at Alamut and in the South Caspian region following the Mongol Conquests. *Journal of the American Oriental Society* 123(2): 351–370.
Voss, Don L. 1958. Ibn al-Haytham's 'Doubts Concerning Ptolemy': A translation and commentary. Ph.D. dissertation. Chicago: University of Chicago.
Walbridge, J. 1983. The philosophy of Qutb al-Din Shirazi; A study in the integration of Islamic philosophy. Ph.D. thesis, Harvard.

Walbridge, J. 1992. *The science of mystic lights: Quṭb al-Dīn Shīrāzī and the illuminationist tradition in Islamic Philosophy*. Cambridge, MA: Distributed for the Center for Middle Eastern Studies of Harvard University by Harvard University Press.

Waṣṣāf al-Ḥazrat, 'Abd Allāh ibn Faẓl Allāh. 1967. *Taḥrīr-i tārīkh-i Waṣṣāf*. Tehran: Bunyād-i Farhang-i Īrān.

Waṣṣāf al-Ḥazrat, 'Abd Allāh ibn Faẓl Allāh. 2010. *Geschichte Wassaf's*. Wien: Verlag der Österreichischen Akademie der Wissenschaften.

Wiedemann, E. 1970. *Aufsätze zur Arabischen Wissenschaftasgeschichte*. Hildesheim: G. Olms.

Wiedemann, E. 2010. "Ḳuṭb al- Dīn Shīrāzī, Maḥmūd b. Mas'ūd b. Muṣliḥ." *Encyclopaedia of Islam*, 2nd ed. Brill Online, 2010. http://www.brillonline.nl/subscriber/entry?entry=islam_SIM-4581.

Wolff, R.L. 1962. The Latin empire of Constantinople, 1204–1261. In *The history of the Crusades*, vol. 2, 187–233. Philadelphia: University of Pennsylvania.

Wüstenfeld, Ferdinand. 1840. *Geschichte Der Arabischen Aerzte und Naturforscher*. Göttingen: Vandenhoeck und Ruprecht.

Yāfi'ī, 'Abd Allāh ibn As'ad. 1997. *Mir'āt al-jinān wa-'ibrat al-yaqẓān fī ma'rifat mā yu'tabar min ḥawādith al-zamān*, 1st ed. Bayrūt: Dār al-Kutub al-'Ilmīyah.

Yakubovskii, A. Yu. 2010. "Marwal- SHāhidjān." *Encyclopaedia of Islam*, 2nd ed. Brill Online, 2010. http://www.brillonline.nl/subscriber/entry?entry=islam_SIM-4978.

Ẓahīr al-Dīn Mar'ashī. 1850. *Geschichte von Tabaristan, Rujan und Masanderan*. St. Petersburg: Kaiserliche Akademie der Wissenschaften.

Zahrawī, Abū al-Qāsim Khalaf ibn 'Abbās al-. 1973. *Albucasis on surgery and instruments*, eds. M. S. Spink and G. L. Lewis. Berkeley: University of California Press.

Zambaur, Eduard Karl Max von. 1955. *Manuel de Généalogie et de Chronologie pour l'histoire de l'Islam*. Bad Pyrmont: Orientbuchhandlung Heinz Lafaire.

Index

K. Niazi, *Quṭb al-Dīn Shīrāzī and the Configuration of the Heavens: A Comparison of Texts and Models*, Archimedes 35, DOI 10.1007/978-94-007-6999-1,

S

T

U

W

Y

Z

MIX
Papier aus verantwortungsvollen Quellen
Paper from responsible sources
FSC® C105338

If you have any concerns about our products,
you can contact us on
ProductSafety@springernature.com

In case Publisher is established outside the EU,
the EU authorized representative is:
Springer Nature Customer Service Center GmbH
Europaplatz 3, 69115 Heidelberg, Germany

Printed by Libri Plureos GmbH
in Hamburg, Germany